Coastal management:
integrating science, engineering and management

INSTITUTION OF CIVIL ENGINEERS

Coastal management:
integrating science, engineering and management

Proceedings of the international conference organized by the Institution of Civil Engineers and held in Bristol, UK, on 22–23 September 1999

Edited by C. A. Fleming

T Thomas Telford

This conference was organised by Thomas Telford Conferences on behalf of the Maritime Board of the Institution of Civil Engineers.

Organizing Committee:
C.A. Fleming, *Halcrow Group Ltd*, Chairman
R. Coninx, *Thomas Telford Conferences*
S. Huntington, *HR Wallingford Ltd*
P. Lane, *Redcar & Cleveland Borough Council*
J. Pethick, *University of Newcastle upon Tyne*
R. Purnell, *Ministry of Agriculture, Fisheries and Food*
I.H. Townend, *ABP Research and Consultancy Ltd*

Co-sponsored by
The Chartered Institution of Water and Environmental Management
Countryside Council for Wales
English Nature
Environment Agency
Ministry of Agriculture, Fisheries and Food
Rijkswaterstaat
Royal Institution of Chartered Surveyors

Cover photograph courtesy of GlaxoWellcome

Published by Thomas Telford Publishing, Thomas Telford Ltd, 1 Heron Quay, London E14 4JD.
URL: http://www.t-telford.co.uk

First published 2000

Distributors for Thomas Telford books are
USA: ASCE Press, 1801 Alexander Bell Drive, Reston, VA 20191-4400, USA
Japan: Maruzen Co. Ltd, Book Department, 3–10 Nihonbashi 2-chome, Chuo-ku, Tokyo 103
Australia: DA Books and Journals, 648 Whitehorse Road, Mitcham 3132, Victoria

A catalogue record for this book is available from the British Library

Availability: Unrestricted
Content: Collected papers
Status: Refereed
User: Civil engineers, local authorities

ISBN: 0 7277 2848 2

© The Institution of Civil Engineers, unless otherwise stated, 2000

All rights, including translation, reserved. Except for fair copying, no part of this publication may be reproduced, stored in a retrieval system or transmitted in any form or by any means, electronic, mechanical, photocopying or otherwise, without the prior written permission of the Books Publisher, Thomas Telford Publishing, Thomas Telford Ltd, 1 Heron Quay, London E14 4JD.

This book is published on the understanding that the authors are solely responsible for the statements made and opinions expressed in it and that its publication does not necessarily imply that such statements and/or opinions are or reflect the views or opinions of the publishers or of the Institution of Civil Engineers.

Printed and bound in Great Britain by Bookcraft (Bath)

Preface

These proceedings contain papers that were presented together with discussion that ensued at the fourth in the series of conferences focussing on coastal management issues, organized by the Institution of Civil Engineers.

The volume includes the keynote paper by Lord de Ramsay of the Environment Agency in which he explores many of the issues facing coastal managers today. The technical sessions cover a variety of highly relevant topics, presented by authors from a range of professional backgrounds.

An international perspective is provided through papers dealing with integrated coastal management projects and regulatory issues in Sri Lanka, the UAE, Barbados, Japan and Denmark. Further topics cover the prediction and management of change, various aspects of recently developed or ongoing coastal management plans and the increasingly important subjects of managing risk and uncertainty.

The final section focusses on themes that are of particular interest in the future and address the need or otherwise for a European directive for coastal management as well as the latest thinking in relation to the UK's flood defence aims, objectives and targets.

In summary, this volume will provide the reader with the latest thinking on a subject which has been rapidly evolving over the past decade.

Professor C. A. Fleming
Editor and Chairman of the Organizing Committee

Contents

Keynote address
    Lord de Ramsey   1

Comparative coastal management

1. **Planning and implementation of coastal zone management in Sri Lanka**
   S. S. L Hettiarachchi   5

2. **Coastal management issues for the Jumeirah Coastline, Dubai**
   J. D. Pos, S. W. Young and Y. Eldeberky   18

3. **The Barbados Atlantic Coast Plan**
   P. W. J. Barter, S. Gubbay and L. Brewster   28

4. **Economic evaluation of coastal management in Japan**
   Y. Imamura   39

Prediction and management of change

5. **Can land consolidation schemes used in Denmark provide a mechanism to facilitate managed realignment in the UK?**
   D. E. Johnson   49

6. **Coping with dynamic change on the coast – do we have the right regulatory system?**
   S. A. John and R. N. Leafe   58

7. **A sea defence strategy for Salthouse on the north Norfolk coast**
   S. J. Hayman   66

8. **Analysis of the long-term variations in offshore sandbanks**
   B. Li, D. E. Reeve and N. Thurston   77

9. **Predicting long-term coastal morphology**
   H. N. Southgate, A. H. Brampton and B. Lopez de San Roman   87

10. **The contribution of technology in supporting data exchange in integrated coastal zone management**
    K. Millard and A. Brady   97

11. **Maximising the use and exchange of coastal data: a guide to best practice**
    P. Sayers, K. Millard and D. Leggett   107

Management plans

12	**The development of a framework for estuary shoreline management plans** N. I. Pontee and I. H. Townend	117
13	**Montrose Bay – coastal management in practice** M. B. Mannion	127
14	**The legal and geomorphic impacts of engineering decisions on integrated coastal management** D. J. McGlashan and G. R. Fisher	137
15	**Taking an integrated approach towards flood defence planning in estuaries** R. A. Cottle, J. G. L Guthrie, N. Pettitt and W. Roberts	147
16	**Shoreline management – a strategic management plan for Scarborough town** J. Riby	156
17	**The Bristol Channel Marine Aggregates Resources and Constraints Research Project: keeping the consultation process clear of murky waters!** J. Brooke	168

Risk

18	**The changing flood risk requirements and the managed retreat policy – Dungeness nuclear power station and how it might be applied elsewhere** R. J. Maddrell and B. Osmond	174
19	**Pitfalls on the path to risk-based coastal management, and how to avoid them** J. W. Hall	189
20	**Managing risk and uncertainty in the design and construction of the Minehead sea defences** M. S. West, M. Caporilli and P. Sedgwick	199

Into the next millennium

21	**Does coastal management require a European directive? The advantages and disadvantages of a non-statutory approach** S. Jewell, H. Roberts and R. McInnes	209
22	**Tourism and resort action plans – identifying a methodology** P. Lane	219
23	**Flood and coastal defence aims, objectives and targets** R. G. Purnell and B. D. Richardson	228

Discussion on papers — 233

Keynote Address

LORD DE RAMSEY, ENVIRONMENT AGENCY

It is particularly pleasing for me to be here, not just because of the timeliness of the subject matter or because of my involvement with the Environment Agency but because it is also quite personal. I live not far from the east coast and I farm below sea level, which concentrates the mind. Also my family have been involved in flood and coastal defence since the seventeenth century. Being environmentally responsible, I, of course, came by train and I walked out of the station to find a large sign which said 'The Reckless Engineer' - I hoped that it was not an omen for this conference.

Nineteen forty-seven was a momentous year for me for two reasons. First because I was taken, at the age of five to the river bank at Earith to see the '47 floods; those were mainly inland rather than coastal. All I can remember is standing on this bank and seeing a landscape that I knew well because we farmed there, and instead of rich black soil and farmyards there was just water, grey water with chimneys and trees sticking out of it. That is the sort of thing, even at the age of five, that stays with you. Second, it was an important year because in July we went on our first holiday after the war to Blakeney on the north Norfolk coast. It was a time when, of course, there was not only food rationing but also petrol rationing and although we had enough coupons to get us to the coast, we had to bicycle from there onwards and anybody who tells you that Norfolk is flat has never been there on a bicycle. We sailed there; we pulled the wrong ropes and were shouted at by my father; we caught flat fish. We went to Blakeney Point and we watched the terns and the seals and we had a wonderful holiday. But, of course, at that age I had no idea of the history of the place and it is something that, interestingly, Stephen Hayman picks up in his paper.

Blakeney was one of three little villages which together formed a major port in the Middle Ages. So important was Blakeney that, when the Spanish invasion was threatened in the 1580s, parts of the coast there were fortified against a landing. I was always brought up to believe that the ports of Blakeney, Wells and Cromer together contributed more ships to fight the Armada than the Cinque Ports on the south coast. I think actually that is a bit of a myth spread about by people in East Anglia but it does give you an idea of the sort of size and importance of our east coast ports. The port itself was effectively ruined by Sir Henry Calthorpe's flood bank, which was built in 1637 and this resulted, as you will hear this afternoon, in the silting up of the river, the River Glaven, and the port itself was delivered a final insult by being bought by the Lynn & Fakenham Railway in the nineteenth century. I did not realise then what a microcosm of coastal history I was looking at. The rivers and coasts then provided our motorways, and Rotterdam was closer to us in north Norfolk than Edinburgh or London. We have many beautiful coasts and they still have many uses, not just commercial and agricultural. As you know, leisure, tourism, sand and gravel, oil and gas, wildlife and landscape all go up to make our coastline. But the sea can pose a threat as well;

Coastal management: integrating science, engineering and management, Thomas Telford, 2000

many low lying areas of land are at risk from flooding, and many parts of our coast are, as you know, at risk from erosion.

So what is the condition of our coasts today? Are we doing enough to protect them for future generations? Is their use sustainable? I have problems with that word because it is often trotted out but I find it a very slippery concept because it means so many different things to different people. I quite liked John Gummer's definition of sustainable development as not cheating on your children.

Traditionally, the management of flood risk and erosion has involved injecting money at the point of trouble, but recent years have seen the development of more enlightened attitudes and approaches — in particular the strategic approach encouraged by MAFF and the National Assembly of Wales on the development and production of shoreline management plans (SMPs). These have greatly advanced our understanding of natural processes in the coastal area and in particular how coastal cells work and interact.

The current emphasis is on the preparation of more detailed strategies that will help realise the benefits from these plans. What I think is interesting is how their production has brought together scientists, engineers, managers and planners to prepare for the first generation of SMPs for our coast. Significant benefits have come from these different disciplines working together and from the synergy created. Arguably the greatest benefit is from engaging and not just consulting the public, especially those people whose homes and livelihoods will be affected by many of the decisions you make in your professional lives.

A particularly good example of working together has been the role that the Environment Agency has taken in managing the Humber Estuary. Over the past three years the Agency has worked very closely with all interested parties to determine objectives for the estuary. These include not just protection of wildlife but also the local economy and its future growth. Probably the most important aspect of the work is that it has not simply resulted in a set of issues and objectives but in an action plan that clearly sets out the Agency's role in achieving these objectives. The outcome of the work has demonstrated that there is considerable scope for potentially conflicting groups to work together. There is now a recognition that working together in real partnerships demonstrates the benefits from protecting environmental and economic assets in a better way than before. And I am not just saying that because the Environment Agency has told me that. I am saying that because I have visited it on three occasions and because, as it happens, it overlaps with my role as a Crown Commissioner. The Crown Estate owns Sunk Island which is an estate of some 10 000 acres on the Humber Estuary. I must say whoever bought Sunk Island must have been quite an optimist because it is not a name to give you great confidence, is it? But having seen what is actually happening there from both the point of view of the landowner and farmer and of the Agency, I can confidently confirm that the Humber Estuary is a very good example which we can all learn from.

One way of working with nature is the use of soft engineering solutions, and beach nourishment is a good example of that. Any proposals demand a joint approach to design philosophy and risk assessment. Any soft design is, by definition, a balance between opposing systems. In recent years, European legislation, especially the Habitats Directive, has caused us to focus even more closely on habitats with special areas of conservation, SACs, and special protection areas and Ramsar sites. But future coastal defence policies and natural processes will have impacts on these habitats and we will hear more about that during this conference.

Therefore, an overview of the potential nature and extent of habitat changes, losses or gains, is absolutely essential in deciding policy. To this end, the Agency, English Nature and interest groups have completed research and development which gives a broad indication of the nature and scale of potential habitat losses and gains based on a best guessed coastal defence scenario for the next 50 years, and that of course also allows for some sea level rise.

In July of this year (1999) the MAFF Minister, Elliott Morley, launched a consultation on coastal habitat management plans, another acronym which conveniently comes to CHAMPs. Initially, these plans will sit alongside our SMPs and over time become fully integrated into them. The recent announcement of support from the European Commission through the project 'Living with the Sea' (led by English Nature) is particularly welcome.

I do have to say at this stage that I find it rather bizarre that we have permissive powers to protect people and property and legal obligations to protect habitats. I say that because another issue pervading our daily lives is that of climate change. I know that this aspect of uncertainty has long been recognised and allowed for and I am sure we will hear more of the damaging effect of increased storminess, flooding and erosion on natural and human assets along those coasts. In some cases this will lead to environmental opportunities. Management realignment and increased use of soft defences are possible ways to address those issues and if these solution are to be implemented and we are serious about a strategic and sustainable approach to managing the UK coastline, then financial consequences must be addressed openly. We must ask what it will mean to people and their livelihoods. In the long term we will see a strategic approach to existing development on eroding coasts. Will we be brave enough to advocate relocation of whole settlements; and if we do, how will we treat those we have encouraged to live and work on the coast, for the benefit not only of themselves but also of the nation? Where it is being advocated that previous management decisions should be reversed, should we not be giving incentives rather than demands? There is no partnership if only one partner is willing.

As you know, generally there is no compensation to those affected by flooding or erosion, including cases where it is decided not to defend a particular area or undertake management realignment. Is that policy right? The statutory Town and Country Planning Process now has new development pressures, yet much of the planning for integrated coastal zone management and shoreline management planning is non-statutory. We must ensure that the planning system is able to deliver what is expected of it for today and for tomorrow. Much progress has been made towards integrating coastal management, but I would urge a note of caution. Coastal management is a long-term process that cannot be stop—started. Decisions are being made now which will affect the coast for many decades and while scientific understanding is increasing, and will continue to do so, this may take years to develop. In contrast, some areas may enjoy a particularly rapid technological development.

In the absence of complete understanding, our policies must be cautious and flexible. Never before has so much data on the coastal zones been available, but we must use that data to make better decisions. We must not fall into the trap of doing too little, too late. In my opening remarks I painted a picture of the economic worth of our coastline. About 10% of our conservation sites are at or near to sea level. Of all UK manufacturing industry, 40% is to be found along the coastlines and estuaries and around 57% of agricultural Grade I land is found at elevations below 5 m. Some 3% of England and Wales is potentially at risk from flooding and by 2016 the number of households in England is predicted to have increased by some 3.2 million or 17%, placing the coastal zone under even greater development pressure.

The challenges facing us all are considerable. We all know that they must be addressed in partnership. We have learned that over recent years. The EU demonstration programme on integrated coastal zone management which reported earlier this year confirmed that most of the physical problems and conflicts observed in the coastal zones of Europe can be traced to procedural, planning, policy and institutional weaknesses. I congratulate those of you who contributed to that report and played a leading role in that programme. Together your various contributions will help us to achieve sustainable development and that will mean a better quality of life for everyone, not just now but for the generations to come.

Planning and Implementation of Coastal Zone Management in Sri Lanka

DR. S.S.L. HETTIARACHCHI,
Dept. of Civil Engineering, University of Moratuwa, Sri Lanka

BACKGROUND

Sri Lanka is an island state with a coastline of approximately 1600 km and a land area of around 64,000 sq. km. Coastal erosion has been identified as a major natural hazard faced by Sri Lanka for a very long time and its impact had increased at alarming rates due to unplanned human intervention in the coastal zone. Although coastal problems of Sri Lanka have been recognised from the turn of the century it was in the early sixties that growing attention was focused on these problems because they had seriously aggravated (1).

The subject of coast protection has been at different times under the direction of different government departments. The focus during the period 1960 to 1970 was mainly to seek solutions to coastal erosion problems by constructing coast protection works designed to satisfy site-specific requirements in a very local context. However, such an approach contributed towards increased coastal erosion problems and the degradation of the coastal environment. This focused the attention of policy makers and the public on the fragility of this zone and the consequences of unplanned development (2).

When the initial response of seeking engineering solutions to curb immediate coastal erosion problems - by the construction of ad-hoc protective structures - proved to be non-effective, it was realised that there was a need to adopt a coordinated approach (3). This need combined with increased environmental awareness resulted in the acceptance of the requirement for some measure of regulation of development activity within the coastal zone. It was also realised that the legislative and administrative framework which was then in existence was insufficient to meet the demands of effective coastal zone management.

The need to establish a separate government department for the management of the coastal zone was identified in the early sixties. In 1964, a Coast Protection Unit was established in the Colombo Port Commission. The need for specific legislation was also identified during this period. The preparation of an initial draft of a new coast conservation law commenced in 1972 to facilitate the control of coast erosion. A more management oriented approach was recommended in the mid seventies as an outcome of a detailed study of coastal problems in Sri Lanka (4). It identified the necessity to establish a separate government department for coast conservation and a research centre for coastal engineering. The recommendations included the preparation of a Coastal Zone Management Plan and the expansion of the draft coast

Coastal Management: Integrating science, engineering and management, Thomas Telford, London, 2000

conservation law prepared in 1972 to include coastal zone management, planning and coast conservation.

A Coast Conservation Division was established in the Ministry of Fisheries in January 1978. A separate Department of Coast Conservation was subsequently established in January 1984. A Coastal Engineering Research Centre was also established, later being transformed to a public company.

COAST CONSERVATION ACT

The Coast Conservation Act No: 57 of 1981 (5) came into operation on 1st October 1983. This Act decreed the appointment of a Director of Coast Conservation with the following responsibilities.

    1) Administration and implementation of the provisions of the Act.
    2) Formulation and execution of schemes of work for coast conservation within the coastal zone as defined in the Act.
and 3) Conduct research in collaboration with other departments, agencies and institutions for the purpose of coast conservation.

It is important to note that the Coastal Zone is defined in the Coast Conservation Act as that area lying within a limit of three hundred meters landward of the Mean High Water line and a limit of two kilometers seaward of the Mean Low Water Line. In the case of rivers, streams, lagoons or any other body of water connected to the sea, either permanently or periodically, the landward boundary extends to a limit of two kilometres, measured perpendicular to the straight base line drawn between the natural entrance points and includes waters of such rivers, streams and lagoons or any other body of water so connected to the sea. These boundaries are illustrated in Figure 1.

The Coast Conservation Act required the Director of Coast Conservation to have a survey made of the coastal zone and on the basis of the results of the survey to prepare a comprehensive Coastal Zone Management Plan. The Act also established the Coast Conservation Advisory Council with responsibilities to review coastal management problems of significant concern and provide advice to the government. It was intended that the Council would facilitate the coordination of all development activities within the coastal zone and play an important role in resolving conflicts between different resource users.

One of the important aspects of the Act refers to the regulation of development activities in the coastal zone under the direction of the Director of Coast Conservation. The regulation of development activities was implemented by a permit system, the procedure for which is laid down in the Act.

COASTAL ZONE MANAGEMENT PLAN

The legal framework for this plan is provided by the Coast Conservation Act No: 57 of 1981 and the Coast Conservation Act (Amendment) No: 64 of 1988. This plan, which received the approval of the Cabinet of Ministers in April 1990, was prepared with the technical assistance

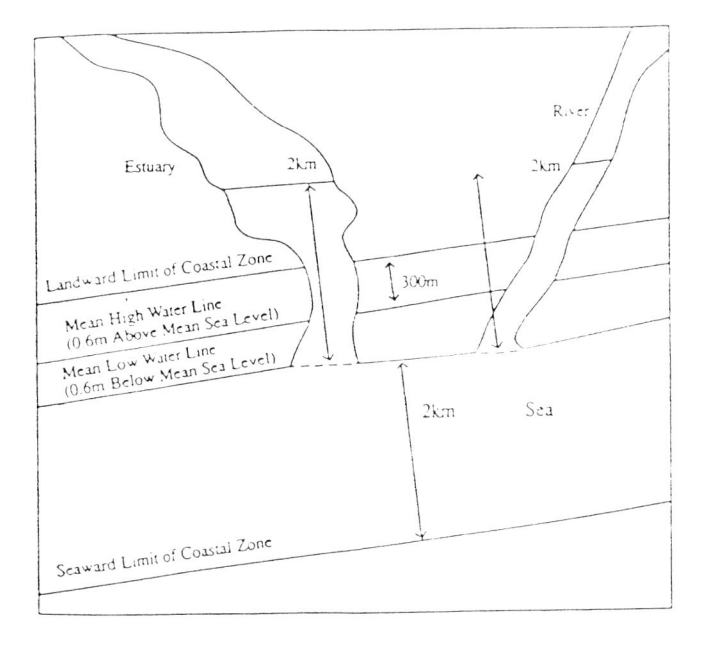

Figure 1 The Coastal Zone of Sri Lanka as defined by the Coast Conservation Act of 1981

of the University of Rhode Island and support of USAID.

The objectives of the Coastal Zone Management Plan (6) are as follows:

- to identify coastal problems that need to be addressed.
- to indicate the reasons why these problems are important.
- to present the management programme adopted by the Coast Conservation Department to address these problems.
- to identify the measures which should be adopted by Government and non-government organisations and the general public to reduce the scope and magnitude of coastal problems and
- to identify research activities of immediate importance to enhance the management of coastal resources.

Within the context of the above objectives the Coastal Zone Management Plan focuses attention on three major concerns, namely, coastal erosion management, conservation of natural coastal habitats and conservation of cultural, religious and historic sites and areas of scenic and recreational value.

The plan describes the nature, scope, severity and causes associated with each of these problems. Objectives and policies for the management of each problem are identified, along with specific management techniques. In particular, the rationale and procedures for continuing the coastal permit system, were outlined in detail.

ENVIRONMENTAL PROBLEMS IN THE COASTAL ZONE

The coastal areas also contain a number of critical ecosystems which include lagoons, estuaries, seagrass beds, mangroves, coral reefs, salt marshes, sand dunes and over 300 km of sandy beaches. Coastal fisheries provide about 80 percent of the total fish catch of the country. These ecosystems provide rich habitats for many forms of aquatic life and they also form a buffer zone protecting land from incursion by the sea. During the last three decades there has been increasing pressure for development in the coastal zone and communities have exploited the use of natural resources such as sand and coral on a commercial basis. Development pressures have also led to the ad-hoc reclamation of estuarial, lagoon and marsh waters and unrestricted disposal of untreated sewage leading to pollution problems (7). In this context there is a need to understand, within the framework of the Coastal Zone Management Plan, the existing problems as well as those which may arise as a result of future development activities.

The important environmental problems in Sri Lanka's coastal zone can be summarised as follows.
1. Coastal erosion, flooding and drainage
2. Unplanned human settlements
3. Construction of maritime structures which interact positively with the coastal waterfront.
4. Construction of buildings and infrastructure within the dynamic coastal zone.
5. Clearing of coastal vegetation.
6. Coral mining
7. Sand mining
8. Discharge of sewage and wastewater

9. Discharge of industrial effluent
10. Agricultural run-off
11. Effects of sand bar formation across river and lagoon outlets, salt water intrusion and drainage.
12. Reclamation of mangrove forests and coastal wetlands
13. Dredging of river estuaries
14. Siltation of lagoons and coastal aquatic systems.
15. Discharge from ships.
16. Oil slicks
17. Breaking of the reef for navigation.
18. Overfishing and collection of exotic fish.
19. Fish farming.
20. Dynamiting and use of explosives for fishing.
21. Pollution of beaches and coastal aquatic bodies
22. Threatened wildlife

It is evident that the above problems had to be given due consideration in developing a management plan for the coastal zone and monitoring its effectiveness during implementation. It was also clear that in view of the relationships which existed between many of the above issues, as well as their impacts on the environment there was a need to adopt a coordinated approach in managing the natural resources and development activities of the coastal zone.

COASTAL EROSION MANAGEMENT PLAN AND COAST PROTECTION

Coastal erosion has apparent and immediate consequences to man and society and is therefore often a principal concern in the administration of coastal districts. During the period 1983-84 the southwest coast of Sri Lanka was subjected to very severe erosion. It was appreciated that any major coast protection work should form an integral part of an overall Coastal Erosion Management Plan and/or a Coastal Protection Plan and if this approach was not adopted it would facilitate the transfer of the problem from one location to another. In view of the severity of the erosion problem foreign funding was sought to arrest the situation. With the assistance of the Danish International Development Agency (DANIDA) the Coast Conservation Department embarked on the preparation of a Master Plan for Coastal Erosion Management, which was completed very successfully in 1986 (8,9).

The work which was carried out under this project included the following areas.

1) Compilation, organisation and synthesis of available information required for the understanding of coastal processes and the identification of critical areas prone to erosion.
2) Recommendation and initiation of such general investigations and studies necessary to complete the existing information network.
3) Formulation of conceptual designs for coastal works in problem areas and recommendations for approaches toward their implementation.
4) Cost assessment of the protection works recommended and preparation of a time schedule for its implementation giving due consideration to sources of funding and resources available for construction.

A careful analysis of previously available and new information led to the preparation of a quantitative assessment of long term erosion trends in Sri Lanka providing a significant input to the Coastal Zone Management Plan. Figure 2 illustrates the coastal development trends from that study.

During the preparation of the Coastal Erosion Management Plan it was evident that the need for coastal protection schemes in Sri Lanka was associated with one of the following categories:

1) Protection of a receding coastline endangering land and other assets of the urban community.
2) Protection of low lying areas under natural protection of a dune system, beach barrier or a nearshore coral reef.
3) Control of undesired fluctuations of the coastal profile around tidal inlets and river estuaries.
4) Maintenance of coastal areas of recreational value
5) Specific protection around and in the vicinity of coastal installations such as harbours, marine highways and rail routes.

The Coast Erosion Master Plan categorized the coastal erosion problem under 'singular cases' and 'key areas'. Singular areas are those with isolated problems covering a short stretch of coastline, mostly resulting from local activity. Key areas refer to a stretch of coastline where the existing development pressures necessitate solutions covering several kilometers. It was expected that such areas require extensive investigations which consider the impact of interference with coastal processes and issues of coastal morphological complexity.

On the completion of the Master Plan for Coastal Erosion Management, DANIDA provided assistance in two stages (1987-89 and 1990-92) for the construction of coast protection works and beach nourishment.

The establishment of coastal cells and the computation of alongshore and onshore/offshore sediment transport can provide valuable information on the local balance or otherwise in the sediment budget. A preliminary study of sediment transport for the southwest coast of Sri Lanka based on coastal cells was undertaken as part of a technical assistance programme provided by the German Technical Assistance Agency (GTZ) for strengthening the Coast Conservation Department (10). It is recommended that this framework be used as the basis for future studies in planning and designing of coastal protection works and also in the evaluation of the impact of such works on the environment.

MANAGEMENT OF DEVELOPMENT ACTIVITIES IN THE COASTAL ZONE
Regulation of development activity and the Issue of permits

The control and management of development activities constitute a major area in the implementation of the Coastal Zone Management Plan. 'Development Activity' is defined in the Coast Conservation Act as an activity likely to alter the physical nature of the coastal zone in any way and includes the construction of buildings and other structures, the deposition of waste or other materials from outfalls, vessels or by other means, the removal of sand, seashells, natural vegetation, sea grass or other substances, dredging, filling, land reclamation and mining or

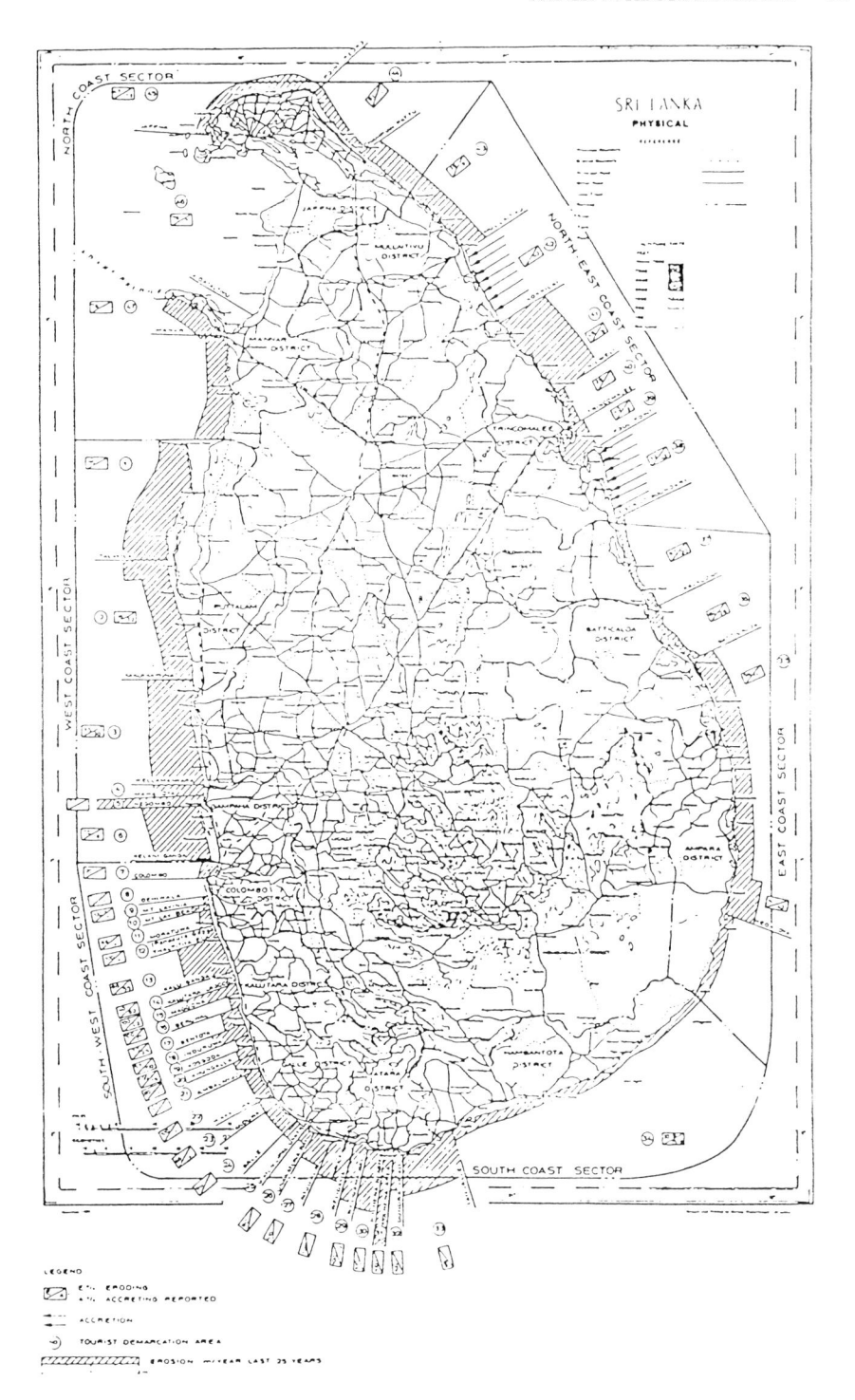

Figure 2   Coastal development trends of Sri Lanka
         (from the Master Plan for Coast Erosion   Management)

drilling for minerals.

The principal means of regulation is via the appraisal of proposed development activities in the designated coastal zone by the staff of the Coast Conservation Department prior to the issue or refusal of a permit to proceed. Although a permit is required for development activities that are likely to alter the physical nature of the coastal zone, it is important to note that fishing, cultivation of crops and planting of trees or other forms of vegetation may be carried out in the coastal zone without a permit.The procedure for the issue of permits is laid down in the Act and is handled by the Coastal Resources Development and Planning Division of the Department.

The Coast Conservation (Amendment) Act No: 64 of 1988 prohibits engaging in mining, collecting, possessing, processing, storing, burning and transporting in any form whatsoever of coral within the coastal zone. The ban on coral mining remains a difficult issue to implement without identifying and making available alternative means of livelihood for the coral miners who have been engaged in this activity for decades. Active governmental support is required to resolve the socio-economic aspects of this problem by generating attractive employment opportunities in the same locality. If such opportunities are created elsewhere due consideration has to be given to problems arising from re-settlement issues.

Sand mining has been controlled to a satisfactory degree by the Department and this activity is allowed only in selected areas of the coastal zone on certain days of the week and to an extent supervised by the staff. In other areas it is a prohibited activity. Nevertheless, violations and infringements of regulation relating to sand mining do occur in some areas and, again, due consideration has to be given to resolving the socio-economic aspects of the problem.

Setback and Variance
An important aspect of the Coastal Zone Management Plan is the 'setback' guidelines which have been identified with due consideration to coastal erosion rates, beach dynamics, shoreline ecology and related activities. A setback, defined as an area left free of any physical modification, is desirable to allow for dynamics of seasonal and long terms fluctuations of the coastline and to ensure pubic access to the water front and visual access to it. The setback standards for construction activities listed in the Coastal Zone Management Plan are designated as minimum standards. These standards are laid down for dwelling units, commercial and industrial buildings, non water-dependent activities and tourist development for different coastal zones as identified in the Plan. Setbacks for water-dependent activities such as hatcheries for aquaculture, boatyards etc. are determined on an individual basis. Setbacks are an effective and inexpensive approach towards coast protection.

Setback guidelines have been introduced to ensure the protection and security of assets which otherwise may be in grave danger due to unfavourable environmental conditions resulting from coastal erosion and tidal flooding. Infrastructure built prior to the Coast Conservation Act of 1981 are not within the setback guidelines and have been very severely affected by coastal erosion. Variance from a standard may be granted by the Director of the Coast Conservation Department only if the Coast Conservation Advisory Council determines that there are compelling reasons for allowing a variance and recommends the granting of the same.

Environmental Impact Assessment

With the introduction of regulations relating to the assessment of the environmental impact of development projects, an Environmental Impact Assessment (EIA) is required in the case of development activities that are considered to have significant impacts on the coastal environment. The Terms of Reference for the study is prepared by the Coast Conservation Department in consultation with the Central Environmental Authority (CEA) and the developer has to prepare the EIA report in accordance with the stipulated Terms of Reference. The EIA report is reviewed by the Coast Conservation Advisory Council and provisions are made for the general public to inspect the same and forward their comments. The Director will give due consideration to the comments received and make a decision on whether to grant approval for the project and on the issue of the permit. EIA reports have been required for a number of development projects ranging from hotel complexes, fishery harbours, land reclamations, coastal highways and coal fired thermal power plant projects.

Contravention of the Act

Engaging in any development activity prior to obtaining a permit issued by the Director and/or non compliance with conditions stipulated in the permit are contraventions of the provisions of the Act and penalties are specified in the Act for such activities. Penalties may include fines and imprisonment and/or confiscation of equipment and machinery used for such activities and/or demolishing of unauthorised structures. It is noted that the Department has demolished unauthorized structures.

Monitoring of Projects

The Coast Conservation Department has responsibility to ensure compliance with the conditions stipulated in the permit through a monitoring system. Monitoring is achieved via periodic site visits or, on some occasions, by direct supervision undertaken by the staff of the Department. The Department may also nominate a state authority for this purpose and sometimes survey teams from institutions such as universities are hired for periodic surveys, for example, to provide coastal status reports on an annual basis, leading to the monitoring of compliance with permits. In the case of large development projects, well- formulated monitoring programmes are incorporated in the EIA report and usually a monitoring committee is appointed to supervise all specified surveys, tests and field investigations.

Decentralization of the permit procedure

The Coast Conservation Department has decentralized several of its functions to the Divisional Secretaries under the terms of Public Administration circular No 21/92 dated 21st May 1992. The delegation of administrative authority has been made under Section 5 of the Coast Conservation Act No. 57 of 1981. This delegation of authority has been formulated to improve the efficiency of the coastal management programme by permitting the local authority to issue permits for restricted sand mining and for construction of small buildings.

Environmental Quality Standards for coastal waters

From the available data on the water quality of coastal waters it was found that, at least in some areas, the level of pollution was unacceptably high, thus threatening the current use. The source of this pollution is mostly disposed domestic and industrial waste water. To protect the users of coastal water as well as the natural habitat it was necessary that water quality standards be established relating to different uses.

In the early nineties, Environmental Quality Standards for surface water were developed by the Central Environmental Authority (CEA) with the technical assistance of the Government of Netherlands (11). A classification for water use was proposed for Sri Lanka on the basis of the present uses and the requirement for water quality. Water quality standards, based on quality requirements, present water quality and tolerance limits in the Sri Lankan context, were assigned to each of the four proposed use classes.

To prevent pollution in sensitive areas of coastal waters, it is necessary to classify them into different zones, each having its own designated use and related quality standards. By doing so it is possible to direct activities which tend to contribute towards pollution to locations where they are least harmful to human health as well as to nature. This classification would also assist in the revision of related quality standards. In this respect broad guidelines were defined for the designation of use in certain zones.

RECENT DEVELOPMENTS IN COASTAL ZONE MANAGEMENT

The first generation efforts in coastal management in Sri Lanka has had many follow up activities including the preparation of Coastal 2000 : A Resource Management Strategy for Sri Lanka's Coastal Region (12). This document focuses attention on specific problems relating to the implementation of the Coast Conservation Act and recommendations are made. These include

- Single agency and sectoral approaches to solving coastal resources management problems to be replaced by a more comprehensive perspective approach.
- The reduction on the emphasis of regulation needs.
- The recognition of the interrelationships among important resource management concerns such as water quality, habitat degradation, use of natural resources and institutional weaknesses and the need to adopt effective strategies involving more than one agency and a range of management techniques.
- The inadequacy of the narrow geographic definition of the coastal zone in the Coast Conservation Act which does not reflect in actual terms the interconnections between coastal ecosystems and resources. This definition has proved inadequate for even the basic tasks of effectively managing coastal erosion and shorefront construction. It is totally inadequate for integrated land - use and water - use plans for coastal ecosystems or habitat management.
- The increase of participation by local and provincial officials and coastal communities in the formulation of plans and strategies for managing coastal resources.

Coastal 2000 provides a strategic view of coastal problems in Sri Lanka and recommends the solutions which need to be tested and implemented over the following decade. It does so in the context of the country's coastal region which it defines as 67 administrative divisions and comprises approximately 24 percent of the total land area, 32 percent of the population, 65 percent of urbanized areas and two thirds of all industrial production.

The Coastal Conservation Department and the Central Environmental Authority have implemented special management projects in order to investigate critical problem areas. The following projects are examples of such positive initiatives.

- Special Area Management (SAM) Projects to study in detail the problems relating to specific areas which are under severe development pressure.
- National Sand Study and follow up studies to investigate the ways and means of preventing and mitigating environmental hazards and degradation caused by present practices of sand mining.
- Environmental Quality Standards for surface water and air quality. These standards are being used to evaluate the present environmental quality, to assess environmental impacts due to discharges and to prepare environmental management plans.
- Wetland Conservation Project to assist in conservation and management of Sri Lanka's wetlands, including several coastal wetlands.

The Coast Conservation Department with the assistance of the Coastal Resources Management Project (CRMP) has reviewed the Coastal Zone Management Plan and revised the main objectives and policies leading to the preparation of a revised plan (13). The revised plan has taken into account the wide experience gained through the first plan and addresses in greater detail issues which had not received such attention earlier. The second generation plan has also given due consideration to the current and projected development trends in the country in refining policies and guidelines.

This plan like its predecessor of 1990, outlines interventions to reduce coastal erosion which may also increase by sea level rise, to minimize depletion and degradation of coastal habitats, and to minimize loss and degradation of sites of archaeological, historical, cultural, recreational and scenic interests. Due recognition has also been given to Coastal Pollution Control and Special Area Management.

CONCLUDING REMARKS

This paper has highlighted important issues relating to the planning and implementation of coastal zone management in Sri Lanka. It has identified the need which existed for coastal zone management to be introduced as well as its different stages of development and implementation. The priority activities undertaken by the Coast Conservation Department with respect to identifying the extent of erosion, the construction of coast protection works and overall coastal zone management planning are outlined. The paper also focuses attention on the management of development activities in the context of regulation, administration and monitoring. Recent developments in coastal zone management have also been summarised.

It is recognized that environmental degradation of the coastal zone is a major hazard faced by Sri Lanka. During the last five years there has been rapidly increasing pressure for development in the coastal zone. In spite of the regulation of development activities, pressure from human activities combined with unfavourable natural conditions have continued to impose adverse impacts on the coastal zone- which provides a range of opportunities for economic development in a wide field of activities.

The experience gained in coastal zone management in Sri Lanka in the last decade has shown that an approach to resource management that focuses on regulation alone tends to alienate the coastal residents affected. It indicates that a collaborative effort on the part of governmental agencies, non-governmental organisations and the local community is required to address the root causes of environmental degradation in the coastal zone. Experience has clearly illustrated that local communities can be organised to manage their natural resources only it they perceive that they will derive tangible benefits from such management and therefore there is need to adopt management policies which can accommodate such interaction. The Department needs to facilitate locally based planning and implementation efforts.

In the above context there is a need to adopt a well coordinated strategic approach in the management of sustainable multiple uses of the coastal zone. This objective could be achieved by developing an integrated coastal management framework for the environmental protection of the coastal zone, which is relevant for three primary reasons. Firstly, because of the interrelationship between coastal zone activities (in particular land use), coastal and flood protection measures and the routine management and control of the coastal zone; secondly, due to widely varying disciplines and techniques that are involved in the analysis of coastal zone problems and their potential solutions; and thirdly, in recognition of the need for spatial integration given the potential physical interaction between neighbouring coastal sections. The present narrow definition of the coastal zone in the geographic context is inadequate and require re-definition for effective management.

A review of previous studies related to coastal zone management in Sri Lanka shows that results of these studies could be used very effectively to provide a sound foundation for the development of an integrated coastal management framework. The development of such a framework would greatly assist in the overall management of coastal zone activities, strategic planning for coastal regions, assessment of regulation needs and the decision making process regarding the transfer of functions to local and provincial authorities. The Coast Conservation Department will be able to transform itself from a primarily regulatory agency to a service-oriented organization. It will enable the Department to provide the leadership, the coordination, the technical assistance and the training that will be required for the successful implementation of a scientifically based coastal planning and management strategy.

REFERENCES

1. ZEPER, J. Coast Protection and Coastal Resource Development in Ceylon. Netherland Bureau for International Technical Assistance. Ministry of Foreign Affairs. 1960.

2. PARANATHALA, W.R. Success and Failure of Coast Protection Works in Ceylon. Proc. of 6th Int. Conf. on Coastal Engineering. Published by Council on Wave Research and The Engineering Foundation, Florida, U.S.A. 1957.

3. GERRITSEN, F & AMARASINGHE, S.R. Coastal Problems in Sri Lanka. Proc. 15th Int. Conf. on Coastal Engineering (Volume IV) ASCE, Hawaii, U.S.A., 1976.

4. GERRITSEN, F. Coastal Engineering in Sri Lanka. Report on the United Nations Mission. 1974.

5. COAST CONSERVATION ACT NO. 57 OF 1981. Parliament of the Democratic Socialist Republic of Sri Lanka

6. COASTAL ZONE MANAGEMENT PLAN. Coast Conservation Department of Sri Lanka. 1986.

7. WICKRAMARATNE, H.J.M. Environmental Problems in the Coastal Zone. Economic Review, Publication of the Peoples Bank of Sri Lanka, 1985.

8. MASTER PLAN FOR COASTAL EROSION MANAGEMENT. Coast Conservation Department of Sri Lanka. 1986.

9. JACOBSEN, P.R. PERERA, N. & JENSEN, K.B. Master Plan for Coastal Erosion Management. Proc. 2nd. Int. Conf. on Coastal & Port Engineering in Developing Countries, Beijing, China, 1987.

10. FITTSCHEN, T., PERERA, J.A.S.C. & SCHEFFER, H.J. Sediment Transport Study for the Southwest Coast of Sri Lanka. CCD-GTZ Coast Conservation Project, 1992.

11. WIJESOORIYA, W.A.D.D. Application of Environmental Quality Standards. Paper presented at the workshop on Environmental Impact Assessment Methodology. Organised by the CEA , Sri Lanka and NAREPP. Wadduwa, Sri Lanka, October 1994.

12. OLSEN, S., SADACHARAN, S., SAMARAKOON, J.I., WHITE, A., WICKREMARATNE, H.J.M. & WIJERATNE, M.S. Coastal 2000 : Recommendation for A Resource Management Strategy for Sri Lanka's Coastal Region, Volumes I and II. Coastal Conservation Department and the Coastal Resources Management Project, Sri Lanka.

13. REVISED COASTAL ZONE MANAGEMENT PLAN. Coast Conservation Department of Sri Lanka, 1997.

Coastal management issues for the Jumeirah Coastline, Dubai

DR. JOHN D. POS[1], STEPHEN W. YOUNG[1] AND DR. YASSER ELDEBERKY[2]

ABSTRACT

The paper outlines the various studies which were undertaken as part of the development of a Coastal Zone Management Plan (CZMP) for the 32km long Jumeirah coastline of Dubai. The Jumeirah Coastal Zone has the best and most intensively used tourist beaches, a rapidly growing number of high quality tourist hotels, harbours for fishing and water based sport, major parks and a wide range of community facilities serving the local population. With much of the Jumeirah frontage subject to erosion, the Coastal Zone Management Plan seeks to provide a safe, stable and attractive coastline as the essential foundation for associated development and for the recreation of the Emirates' rapidly expanding population.

The paper first reviews the coastal management issues for the frontage and then describes the extensive data collection and field measurement exercise which was undertaken, which included topographic, bathymetric surveys, sediment sampling and wave and current measurements. Much of this data was used as input to the numerical and physical modelling studies.

The paper then describes the extensive numerical modelling of the full study frontage which was undertaken using the MIKE21 and LITPACK packages developed by DHI. LITPACK was used to determine the longshore sediment transport and morphodynamics of the study frontage for both the existing frontage and three coastal defence management strategies. MIKE21 was used to determine detailed wave, current and sediment transport climates adjacent to the major existing coastal structures and harbours along the frontage. The numerical model studies were supported by limited physical model studies at a 1 in 43 scale.

Finally, the paper outlines the coastal development vision of the Coastal Zone Management Plan for the frontage. The various elements of the Plan which support this vision are then described.

1.0 INTRODUCTION

The Dubai population is expected to increase from 637,800 in 1993 to 2,092100 in 2012, a growth of 300% in approximately 20 years. The majority of Dubai's population increase is to be accommodated in the coastal zone. Also tourist visits to Dubai are expected to increase from 12,600 per day in 1998 to 24,400 in 2012, almost doubling in fourteen years. In addition to the rapidly increasing pressure on the coastline from population growth and tourist demand, the coastline is subject to escalating erosion with loss of amenity beaches and damage to property. The Jumeirah frontage of Dubai which covers a 32 km frontage from Port Rashid to the Port of Jebel Ali (approximately 50% of the total Dubai coastline) was recognised as requiring immediate attention. Consequently, the Dubai Municipality commissioned Mouchel to compile

---

[1] Mouchel Consulting Limited, West Hall, Parvis Road, West Byfleet, Surrey, KT14 6EZ, UK

[2] Dubai Municipality, PO Box 67, Dubai, UAE

Coastal Management: Integrating science, engineering and management, Thomas Telford, London, 2000

and implement a Coastal Zone Management Plan for the Jumeirah Coastline. This paper outlines the coastal management issues and major study elements that contributed to the development of the plan and concludes with the description of the main elements of the CZMP focussing on the strategy for the sustainable defence and management of the Jumeirah coastline.

2.0 COASTAL MANAGEMENT ISSUES
2.1 Introduction

Dubai has a Gulf coastline of approximately 64 km. Half of this coastline is within the Jumeirah Coastal Zone. The coast beyond this zone is occupied by existing urban development, ports and designated conservation areas. The most significant opportunities for new coastal development are concentrated in the Jumeirah Coastal Zone which already has the best and most intensively used tourist beaches, a rapidly growing number of high quality tourist hotels, harbours for fishing and water based recreation, major parks and a wide range of community facilities serving the local population. The coast has already been a significant attraction for new development, resulting in the creation of a linear urban corridor from Port Rashid to Jebel Ali Port. The Gulf coastline will continue to be the focus for both international tourism and local leisure facilities. Maintaining a safe, stable and attractive coastline provides an essential foundation for Dubai's expanding tourist economy, and for the comfort and recreation for the Emirate's rapidly growing population.

2.2 Coastal Character

The original character of Dubai's coastline was continuous long sandy beaches, with wide, open views to the sea. This character has been modified by the development that has taken place. Coastal structures including ports, fishing harbours, groynes and breakwaters have interrupted stable beach conditions. Sand accumulation in protected areas of the coast has created deep beaches whilst exposed areas have suffered severe erosion. The Coastal Zone Management Plan is intended to stabilise beach formation and to enhance the character of the coastline. These objectives have strongly influenced the selection of the preferred coastal protection measures.

2.3 Coastal Protection

As part of the Coastal Zone Management Plan study the historical events which have created the current problems of tidal flood risk, severe coastal erosion and sediment loss from parts of the frontage were researched. Modelling techniques, based upon detailed survey data, were used to assess the performance of both existing and planned coastal defence measures. A vocabulary of possible protective structures were identified and evaluated. The selection of appropriate solutions was undertaken on a carefully considered basis for individual lengths of shoreline to match existing coastal conditions, the desired character of the coastline, and in response to onshore development opportunities.

The range of possible protective structures considered include:

– modifications to existing offshore structures to improve their performance
– new shoreline structures in the form of groynes and breakwaters, designed to protect existing beach areas, and to create new shore normal beaches
– offshore breakwaters to provide sheltered areas for beach recreation or safe watersports
– new protective revetments combined with vehicular or pedestrian corniches

The shoreline structures that have been considered have been of both shore normal and shore parallel forms. Alternative structure spacings have also been considered resulting in the opportunity to create a variety of coastal characters.

2.4 Key Issues

A number of key issues have determined the form and content of development proposals:

- the physical integration of coastal protection structures with existing development, and new development opportunities
- increasing the awareness of the coastline and its facilities
- improvement of the character of Al Jumeirah Road through coastal linkage
- enhancing the visibility of the coast
- improving access to the coast for both tourist and local residents, for pedestrians, private cars and public transport
- acknowledging and exploiting the coastal "development nodes" identified in the Dubai Structure Plan
- integrating the "Green Corridors" proposed in the Dubai Structure Plan with plans for traffic circulation, access to the coast and visual corridors to the sea
- improving the overall environment, facilities and comfort for all visitors to the coast
- restoring and maintaining the existing open coastline character of the frontage
- respecting, as far as possible, the visual privacy of existing residents
- stimulation of further investment by the private sector and others.

2.5 Coastal Development

The demand for new development in the coastal zone is based upon the recreational needs of a constantly growing resident population, and a rapidly expanding tourism market.

The projections of population growth, employment structure and the expansion of international tourism emphasise the importance of Dubai's coastline as a major generator of national income. Within the next decade almost half the working population of Dubai will be involved in tourism and leisure related industries, and the construction projects required to satisfy that demand. There were, in 1996, over 230 hotels in Dubai with only five operating in the Coastal Zone. The Coastal Zone is considered to be a major unexploited asset for economic development within the Emirate. Pressure for new development within the Zone is already intense, and opportunities to meet this demand are limited.

The Coastal Zone Management Plan which was developed is based upon a balanced response to demand for new development, and the need to provide safe and stable beaches at acceptable cost, located to produce maximum development potential. Analysis of the Coastal Zone, and selection of appropriate defensive structures have defined major development opportunities, which coincide with the longer term planning objectives of the Dubai Structure Plan.

3.0 FIELD AND MODELLING STUDIES
3.1 Introduction

An extensive programme of data collection, field measurements, numerical and physical modelling was undertaken to facilitate the identification and selection of coastal protection and enhancement options for the frontage. Mouchel, as far as possible, adopted UK best practice developed through their extensive experience of Shoreline and Coastal Zone Management both in the UK and internationally. For example, following the UK Shoreline Management Plan approach, the Dubai coastline was split into 15 Coastal Process Units. These are stretches of coastline which are substantially independent and homogenous as regards coastal processes.

To facilitate the development of planning strategies for the frontage, it was split into 5 planning sectors. The sector demarcations were based primarily on existing and future proposed land use.

3.2 Data Collection and Field Measurement

As an integral part of the study, a comprehensive programme of field studies was commissioned. The field studies provided base bathymetric and topographic charts for the frontage, as input to the other studies. The field measurements also provided calibration and verification data for the numerical and physical modelling studies and allowed parameters to be derived for the design of the beach protection systems. In addition, the field studies provided base data for the development of a long-term monitoring system for the frontage and a Geographic Information System (GIS).

The field study programme included :

- bathymetric survey from Dubai Dry Dock to Jebel Ali Port.
- topographic beach surveys from Dubai Dry Dock to Jebel Ali Port.
- sediment sampling and testing over the survey area
- directional wave measurement at two locations
- tidal recording
- current recording
- inshore non-directional wave measurement
- drogue tracking
- photogrammetric survey and mapping of the entire frontage.

In addition, historical shoreline positions were identified from early aerial photography and tide and wind records were obtained from third parties.

Historical coastline positions were compared for the years 1986, 1989, 1992, 1993 and 1994. For the years 1981 and 1991 digital terrain models were derived for the entire frontage from aerial photography. These were then compared with the 1996 survey undertaken as part of this study to yield detailed information on coastline evolution. Comparison of the digital terrain models has allowed volumes of accretion and erosion and sediment transport rates along the frontage to be established. These have then been used to calibrate the numerical modelling.

3.3 Numerical Modelling

The entire Jumeirah frontage from Dubai Dry Dock to Jebel Ali Port has been modelled numerically. The primary numerical modelling tools that were utilised in this study were the LITPACK and MIKE 21 suites of programmes developed by Danish Hydraulic Institute. A plan showing the coverage of the various models is given in Figure 1. As far as possible the models have been calibrated and verified against field measurements and physical model results.

LITBASE was used to undertake a baseline study of the coastline from Dubai Dry Docks to Umm Suqeim I. The goal of this model was to gain knowledge of the historic coastal evolution and on that basis to establish a sediment balance for the shoreline.

A further baseline study of the area from Jumeirah Beach to the Al Rais Breakwater was undertaken in MIKE 21 E. The purpose of this model was to obtain a detailed appreciation of the wave, current and sediment transport conditions that existed prior to the construction of the beach protection structures in that area.

The LITPACK North model was developed to analyse a range of management options for the frontage from Jumeirah Beach to the Al Rais Breakwater. Three concepts were tested in the model; long groynes, artificial headlands and headlands with revetment north of Jumeirah fishing harbour.

Models MIKE 21 A and B were applied to provide a detailed understanding of the processes occurring in the Jumeirah beach area and around the offshore breakwater. These two models were subsequently combined to form a single model. The models were also used to verify and optimise the alternative management options for the frontage. The predicted wave heights and current patterns were compared with those from the physical model and the currents from the field surveys. Satisfactory agreement between the measurements and models was achieved.

Jumeirah Fishing Harbour, Umm Suqeim I Harbour and Mina Siyahi Harbour were modelled in MIKE 21 C, MIKE 21 F and MIKE 21 D, respectively. The purpose of these models was to determine the wave, current and sediment transport conditions in the vicinity of the harbours for the existing layouts and a series of possible modifications.

The Sheikhs' Palaces' Harbours and Island and the Chicago Beach Resort Marina and Island were similarly modelled in MIKE 21 H and MIKE 21 G respectively. For the Sheikhs Palaces Harbours and Island frontage, the primary focus of the modelling was to control the migration of sand from the Palace beaches to the lee of the island. For the Chicago Beach Resort the primary focus was to investigate the effect of a proposed updrift groyne field on the coastal processes of the resort frontage.

LITPACK Central was established to model the frontage from Umm Suqeim 1 harbour to Mina Siyahi harbour. LITPACK South was established to model the frontage from Jebel Ali Port to Mina Al Siyahi Harbour including the DEWA frontage. The shoreline was modelled with revetment installed over the industrial frontages. LITCOAST was used to predict the coastline evolution in 5, 10, 20 and 30 years relative to the existing coastline. LITCOAST was also used to model the coastline evolution for three generic coastal defence strategies namely:

(i)   Strategy A: High Capital Investment/Low Maintenance Strategy.
(ii) Strategy B: Intermediate Capital Investment/Intermediate Maintenance Strategy.
(iii) Strategy C: Low Capital Investment/High Maintenance Strategy.

LITCOAST was also used to model the preferred solution comprising a modified Strategy A.

3.4 Physical Modelling

To provide calibration and verification of the numerical models in the area of the DMBC Breakwater, T Breakwater and Offshore Breakwater, these areas have also been modelled in a pair of complementary physical models, Figure 1. The model scale was 1 in 43. The models were constructed with fixed beds and sediment movement patterns were simulated using tracer sand on the bed.

PAPER 2 : POS, YOUNG AND ELDEBERKY   23

Jumeirah Coastal Zone Management - Phase 3
Numerical Models and Physical Model Coverage

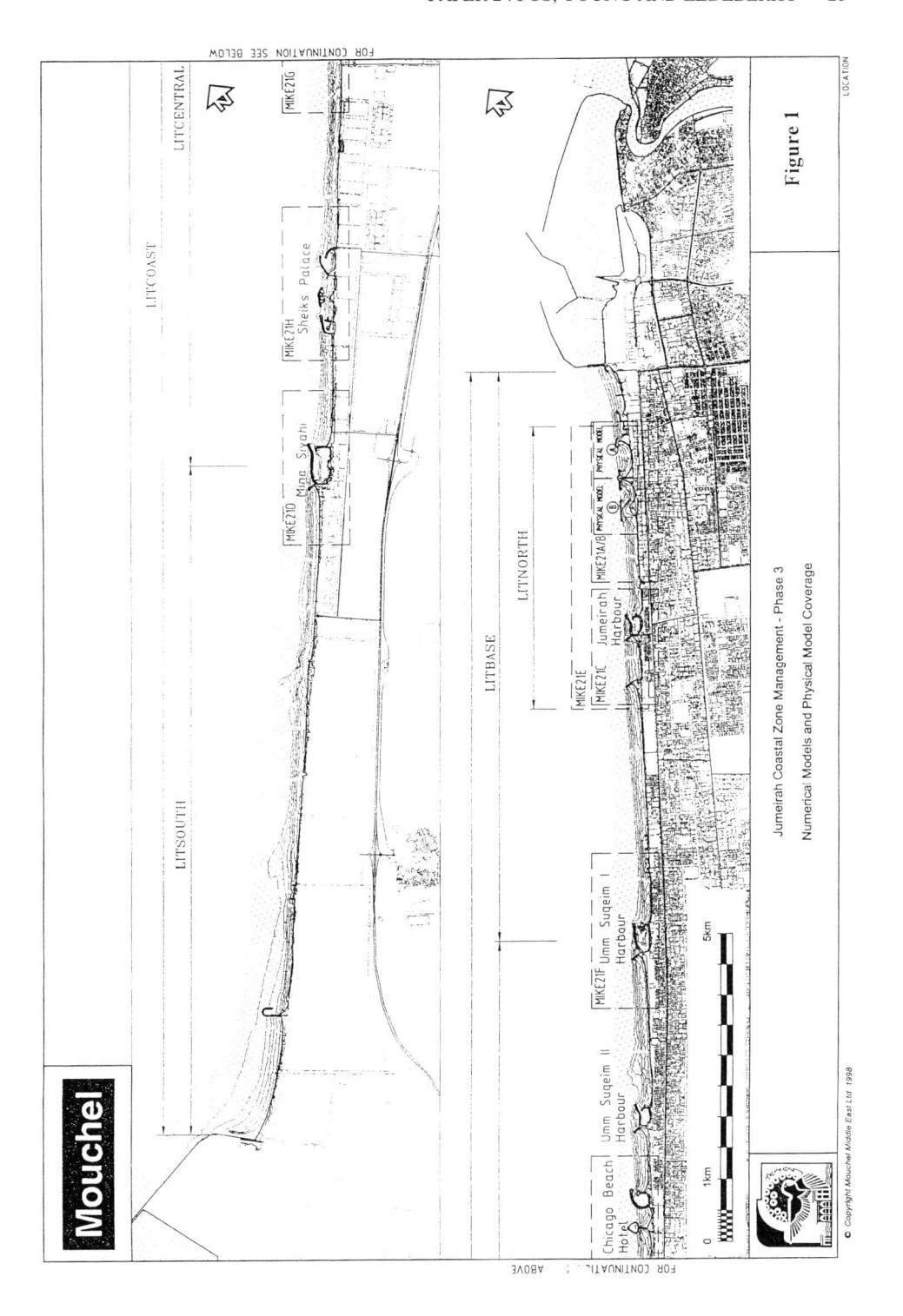

Figure 1

As well as providing data for the calibration and verification of the numerical models, the physical models were also used to assess the performance of different alternative layouts for the coastal management measures to add further support to the numerical modelling studies.

The results from the physical models generally compared well with the numerical modelling analysis providing a high level of confidence in the applicability of the modelling to the conditions experienced on the Jumeirah coastline. Where discrepancies between the models were identified, this was generally able to be explained as a consequence of the limitations of the techniques being followed.

3.5 Environmental Survey

Marine Ecology

The marine ecology up to 2 km seaward of the Jumeirah frontage between Port Rashid and the Port of Jebel Ali was investigated during April 1997. A diving survey was undertaken along 31 shore normal transects extending from the shore to 2 km offshore. The principal aim of the study was to determine the presence of important marine habitats and species which may potentially be affected by proposed coastal engineering and associated works. Sea grasses and coral communities were identified as being of particular importance. In general, seagrass distribution within the study area was found to be concentrated in the southwest, while the distribution of corals in the study area is very limited.

Terrestrial Ecology

Areas of terrestrial conservation value were identified by consultation with local specialists and by site visits.

Cultural Heritage

The study area incorporates sites and features of significant cultural heritage including mosques, a historic building, archaeological sites and traditional ship building yards.

Fisheries

Along the study frontage, fishing is centred at the Jumeirah, Umm Suqeim I and Umm Suqeim II harbours.

4.0 COASTAL ZONE MANAGEMENT PLAN

The Jumeirah Coastal Zone Management Plan presents a vision for the development of the coastal area. This development plan is necessary to meet the demands of rapid population growth, increased tourism and expanding urban development along the coast. Combined with a desire to regenerate and develop the coastline for the benefit of the whole of Dubai, the implementation of the Plan offers tremendous opportunities to improve the recreational and tourism potential of the coastal zone. The Management Plan provides the basis for further more detailed local plans for the key areas and provides a framework to both guide and stimulate future coastal development.

The Coastal Development Strategy is based upon:

- The creation of visible gateways to the coast, where development opportunities coincide with coastal protection and the creation/expansion of beaches. "Gateway" opportunities have been defined at:

- Jumeirah Beach
- Jumeirah Harbour
- the telecommunications site
- Umm Suqeim I and II Harbours
- Chicago Beach Resort Development
- Mina Siyahi Harbour
- South of the SAS Hotel

- Coastal protection structures to enhance and stabilise existing beaches whilst maintaining an open coastal aspect. The proposed coastal protection structures include:

  - extensions to the groynes and breakwaters at Jumeirah Beach and creation of a new intermediate offshore breakwater
  - construction of a new offshore breakwater at Jumeirah Beach
  - extension of the existing Jumeirah Beach corniche to Jumeirah Harbour
  - extension of the updrift spur to Jumeirah Harbour
  - offshore islands to provide calm water sports areas
  - extension of the Al Rais groyne to the north of the Hilton Beach Club
  - construction of a new groyne from the telecommunications site
  - construction of new revetments and a landscaped linear park north of Umm Suqeim I Harbour
  - extensions to the updrift breakwaters to Umm Suqeim I and II Harbours
  - construction of a new groyne to the south of Chicago Beach Village
  - construction of a new groyne to the south of the Royal Palaces
  - extension of the updrift spur to Mina Siyahi Harbour
  - construction of a new groyne to the south of the SAS Hotel

  In addition to the coastal structures listed, there is the opportunity to create additional beach amenity areas. This would be achieved by a further groyne north of Umm Suqeim I Harbour, a groyne north of the Sheikhs' Palaces' harbours and by a string of offshore islands south of Mina Siyahi.

- Nodal development opportunities.

  Through the Coastal Zone Management Plan existing features of the shoreline will be developed and enhanced whilst other new development sites will be created.

  In particular the existing harbours are identified as having tremendous opportunities as focal points that are capable of being enhanced to form additional attractions on the shoreline. The existing structures at Jumeirah Beach corniche are similarly an area where the opportunity exists to extend the usage of the beach amenity.

  Other sites for development along the frontage include Al Rais Breakwater area, Jumeirah Beach Park and the adjacent telecommunications site, the linear frontage north of Umm Suqeim I as well as the area adjacent to the Chicago Beach Resort Development.

- The enhancement of Al Jumeirah Road as a coastal highway.

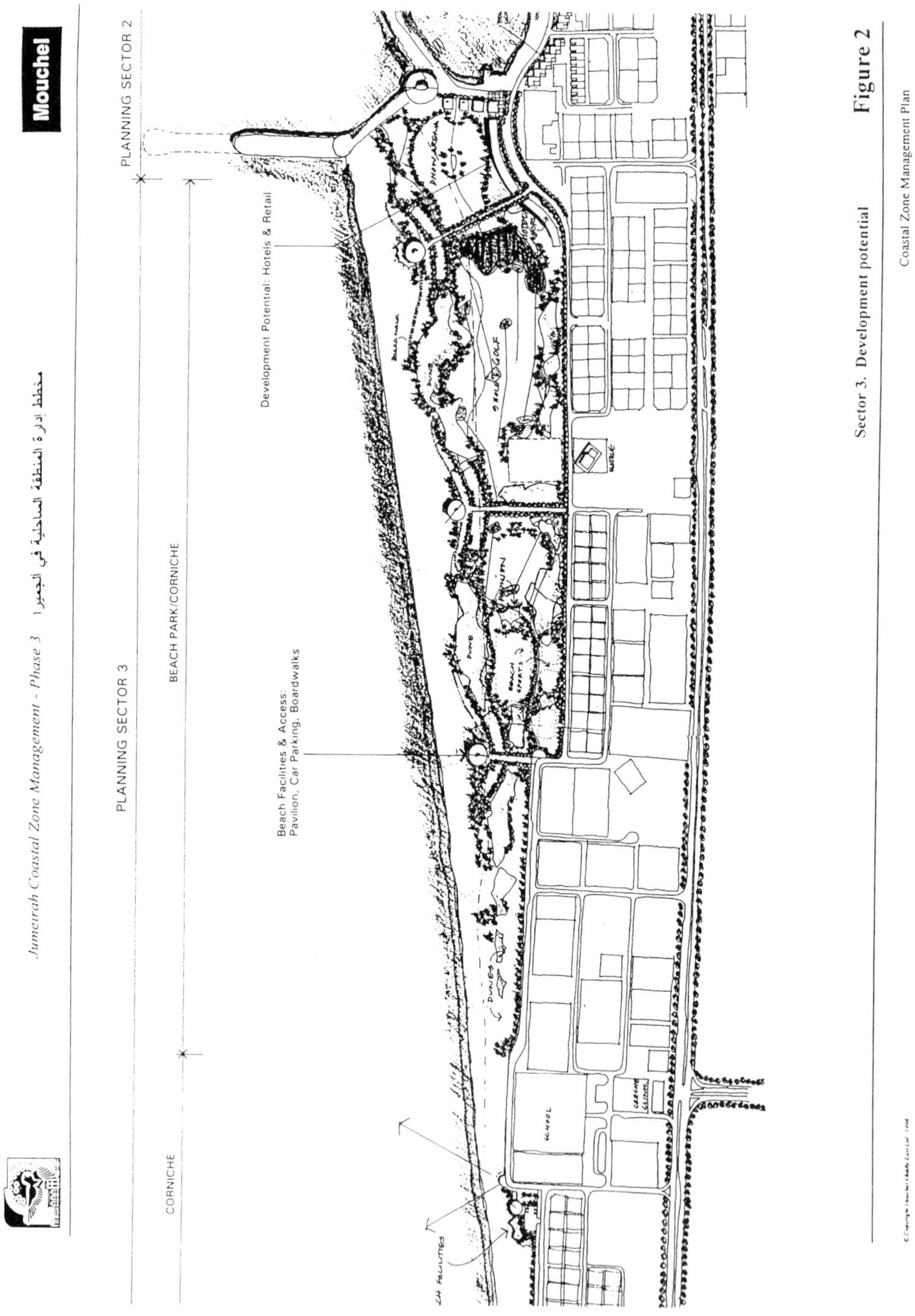

26 COASTAL MANAGEMENT

Figure 2

Sector 3. Development potential

Coastal Zone Management Plan

Al Jumeirah Road is the main arterial highway which provides access to the coast, and intersects with the "green corridors" proposed in the Structure Plan. The existing road reservation should be intensively planted to create a "green coastal corridor" with clearly defined "gateways" to the coast.

- Reclamation for the creation of additional waterside and beach developments.

  The coastal defence strategy will create substantial additional reclamation that will provide future development sites. These substantial areas will facilitate the improvement of the recreational opportunities along the coast through parks and commercial opportunities for further investment. Such sites will be reclaimed adjacent to Al Rais Breakwater, Umm Suqeim I Harbour, Umm Suqeim II Harbour and Mina Siyahi Harbour. As an example, the proposed beach park development to the south of Umm Suqeim I harbour is shown in Figure 2.

- Supporting infrastructure to stimulate development opportunities.

  Through the Coastal Zone Management Plan Contracts the coastal defence works for the frontage will be constructed in a phased manner. In conjunction with these, the basic infrastructure will be installed to include roads, drainage services, parking provision and basic landscaping.

  The provision of this basic infrastructure is considered essential to the stimulation of private sector investment into the coastal frontage. Once this infrastructure and the framework for development of the coastline is in place it is considered that it will form a significant attraction to the private sector to invest in hotels, beach clubs, restaurants, leisure areas and similar features.

5.0 CONCLUSIONS

A comprehensive Coastal Zone Management Plan has been prepared for the Jumeirah Coastline from the Dry Docks to Jebel Ali Port.

The Plan provides for the stabilisation of the coast through a comprehensive programme of coastal engineering works. The principle underlying the works has generally been to achieve stable coastal alignments through reorientation of the shoreline. The study approach was based on UK best practice.

The stabilised shoreline will provide the basis for development of the recreational and amenity potential of the coast to serve the needs of the national and international populations into the future.

The Coastal Zone Management Plan provides the framework for future development of the coastal area. Development control guidelines have been proposed to ensure that future development that is proposed takes place in accordance with the Plan and in a manner that will not impact detrimentally on the coastline and the processes controlling its evolution.

The coast is a tremendous asset to Dubai whose full potential has not been realised due to limited access and erosion of beaches. The Costal Zone Management Plan provides the means to unlock this potential.

The Barbados Atlantic Coast Plan: An effective integration of new and focused scientific field studies with existing works and data records, the ingredients of the island's future sustainable development

P. W. J. BARTER, Engineering Manager, Halcrow Group Ltd, Swindon, UK,
DR S. GUBBAY, Consultant, Ross-on-Wye, UK,
L. BREWSTER, Assistant Director, Coastal Zone Management Unit, Barbados.

INTRODUCTION
The most easterly of the Caribbean islands, Barbados is unlike its volcanic near neighbours along the Lesser Antilles chain in that it has been created by the vertical thickening of sea floor sediments. The emergence of the island above sea level in three pulses over the last 15 million years is observed in its distinctive raised cliffs, a prominent feature of the country's landscape. In terms of its size, 32kms long by 24kms wide, and certain of its physical characteristics it has a lot in common with the Isle of Wight. A map illustrating some geographical locations referred to in the paper is included as Figure 1.

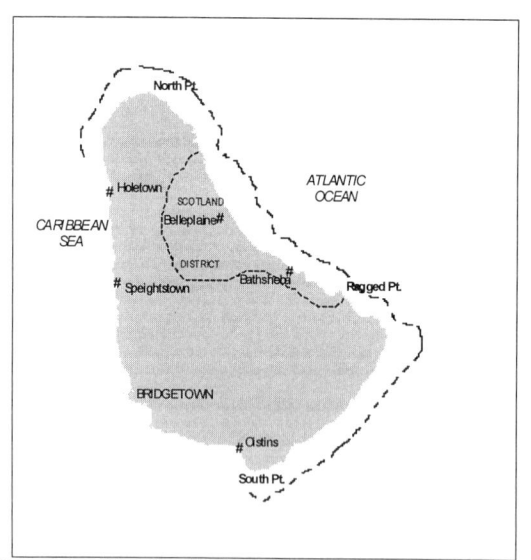

Figure 1 – The Study Area in a Regional Perspective

Its Atlantic seaboard faces the trade winds and is therefore continually exposed to high wave energy. This is reflected in the rugged landscape to the north and along its south-east coasts. In between on this coast is the influence of the Scotland District where erosion has exposed

Coastal Management: Integrating science, engineering and management, Thomas Telford, London, 2000

an ancient set of sands and clays deposited in deep water before the island emerged from the sea. The steep unstable terrain sloping eastwards towards the sea poses severe problems for maintenance of roads and buildings, and for agriculture. Material carried by its intermittently flowing watercourses provides the bulk of the material that form the beaches and dunes that characterise the frontage. Because of its unstable terrain and exposure, residential development between North Point and Ragged Point has in the past been limited. In contrast, a corridor along the relatively flat and wide terrace behind the first cliff along the south-east coast has become the preferred area for new housing development. Along this coast the shoreline and nearshore is influenced by two reef systems. Further north other than local terracing, the seabed slopes steeply away.

The Caribbean Coast is characterised by long narrow sandy beaches and fringing reefs and has been the focus for tourism related development. It also houses some important industrial facilities. These developments which started in the 60's, have been at some environmental cost. This includes loss of coastal wetlands, degraded reef and coastal water quality, encroachment of buildings on active beach areas and loss of beach access.

BACKGROUND TO THE COMMISSION
The Government of Barbados has been developing its capacity to implement coastal management over many years. During the 1960's and 1970's there were scientific research programmes focused on the Caribbean coast of the island, gathering information on aspects such as reef ecology, coastal processes and water quality.

These were developed into systematic studies in the early 1980's and revealed the threats posed by pollution and shoreline instability along these southern and western coasts of Barbados. A Coastal Conservation Project Unit was established to carry out further investigations on these issues and this group, now known as the Coastal Zone Management Unit (CZMU), has recently become an agency within the newly created Ministry of the Environment, Energy and Natural Resources.

The CZMU has been responsible for overseeing two major coastal management projects. The first of these proposed new legal and organisational mechanisms to remedy the problems along the south and west coasts, including two new pieces of legislation recently approved by Parliament - a Coastal Zone Management Act and a Marine Pollution Control Act. That commission also produced a draft Coastal Zone Management Plan for the south and west coasts (DELCAN, 1995a), gave proposals for an investment phase programme aimed at enhancing environmental quality (DELCAN, 1995b), and effected the construction of a number of pilot schemes, aimed primarily at stabilising or enhancing various sections of the shoreline.

The second is the more recent project and the focus of this paper for which a team led by Halcrow was appointed. This has included the preparation of an Integrated Coastal Management (ICM) Plan for the remaining south-east, east and north-west coasts (Halcrow, 1998a), a summary of the national framework for ICM in Barbados (Halcrow, 1998b) and guidance on further requirements to develop the legal and institutional framework to support ICM (Halcrow, 1999a). Following on from these a comprehensive programme of investments associated with the implementation of both Plans has been produced.

FIELD SURVEYS AND DIAGNOSTIC STUDIES

Unlike the Caribbean Coast that has been subject to a wide range of past studies and research, the level of knowledge and understanding about the environments of the Atlantic Coast was found to be limited. It was for this reason that the original terms of reference for the commission had put a strong emphasis on the need for establishing baseline data. Mindful of the difficult on and offshore conditions under which the field teams were to be deployed to collect such data, the project called for the use of a range of different interlinked techniques.

Air photography was used to inform the on-land components and Light Detection and Ranging (LIDAR) (Halcrow, 1997a), the marine programmes. The former campaign provided 1:5,000 photographs, and a mozaic of 2km x 2km imagery for loading on the project Geographic Information System (GIS). The latter survey managed to provide a detailed and comprehensive bathymetry of the seabed out to the 40m isobath with a positional resolution of 3m and an accuracy of +/- 0.3m. Both campaigns were commissioned and successfully completed early in the project. The range of field surveys and studies that were undertaken as part of the programme included the following:

- Mapping and description of all onshore geology – assisted by the aerial photography the project was able to bring together and expand upon the field mapping built up over many visits by American geology experts. The end result is a benchmark geology description of the study area (Halcrow, 1998c, Annex 1).

- Mapping and characterisation of land use – specialist techniques were used to analyse the aerial photography and create digital files, geo-corrected to allow them to be overlain upon existing Lands and Surveys mapping, for a range of land use categories. These were subsequently quality reviewed in the field (Halcrow, 1998d).

- Mapping and description of terrestrial flora – the land use analysis was used by a locally based expert as the backdrop to his own detailed flora mapping (Halcrow, 1998e).

- Assessment and description of all engineering structures – with the assistance of the aerial photography, all engineering structures were identified and their condition was assessed (Halcrow, 1998f).

- Socio-economic surveys to establish the characteristics of coastal communities – local survey teams were deployed within a number of coastal communities to obtain a wide range of data about those communities (Halcrow, 1997b).

- Telephone surveys – these took place at quarterly intervals over a yearly period and were used to identify and assess the importance of the coastal environment to Barbadians. Altogether responses from nearly 2000 households were analysed (Halcrow, 1999b).

- Offshore geophysical and bathymetric survey and interpretation – a field campaign informed by and overlapping with the LIDAR survey. The survey was able to extend the bathymetry out to the 200m isobath. Side-scan and boomer deployments provided information on the nature of the seabed out to 100m depth (Halcrow, 1998c, Annex 2).

- Sediment sampling and analysis – over 600 samples were recovered for sediment trends and carbonate analysis. In conjunction with other modelling work sediment movements were determined (Halcrow, 1998c, Annexes 3 and 4).

- Acoustic Bottom Classification (ABC) – in conjunction with the sediment sampling programme, the ABC provided further information about the likely nature of the seabed (Hacrow, 1998g, Annex 2).

- Marine and terrestrial water sampling and analysis (Halcrow, 1998h) – samples of groundwater, on-land surface water and marine water were taken at intervals over a period of 10 months and analysed at the Government's Analytical Services (GAS) Laboratory. In conjunction with this an exercise to help strengthen the laboratory was undertaken. As well as providing new equipment and establishing quality control and quality assurance procedures, a specialist analyst was seconded to GAS for the initial period of the analysis programme (Halcrow, 1998i).

- Marine habitat mapping and description – the LIDAR provided an initial reference upon which the subsequent diver surveys were based and the ensuing survey data mapped. Later comparison with the findings of the geophysical and ABC surveys allowed for final refinement of the map area boundaries (Halcrow 1998g).

- Oceanographical and meteorological data collection and analysis – a range of available data supplemented by new field collected data provided the basis for a range of modelling work. A short term Ocean Surface Current Radar (OSCR) campaign involving deployments at three different sites provided useful insight into the complex flows around the coast (Halcrow, 1998j).

- Beach profile data collection and analysis – through a programme instigated in 1983, the island now has an extensive data set upon which to draw. Of the 75 beaches currently surveyed, 21 of them fell within the area of this study. Using the Shoreline And Nearshore Data System (SANDS) provided under the project, the data available has been reviewed and analysed (Halcrow 1999c).

- Public consultation and issues analysis – various forms of consultation took place. Key officers within government were consulted on a one-to-one basis. In order to get reaction from the wider public, churches, schools, newspapers, radio and television were used. More direct responses were obtained through the use of a stand at an annual international trade fair and through a road-show that visited a number of sites within the area of study (Halcrow, 1998k).

Examples of some of the data collected and the way it has been presented are included in the selection of slides that are reproduced in this paper. Considering an area on the east coast, Figure 3 features base photography overlain by the digitally held rectified building outlines and roads. Figure 4 illustrates the same area's land use classification extracted from the original photography. Figure 5 shows terrestrial flora classification for that area, developed from the boundaries established by the land use interpretation. Looking to the area immediately offshore, Figure 6 shows the mapping of offshore marine habitats and with it the data resulting from the ABC survey. This series of figures is then completed by data sets that cover both the land and sea environments. Figure 7 features a merging of the offshore geophysical data and the onshore geological mapping and Figure 8 shows a similar merging of onshore topography and offshore bathymetry data sets.

32 COASTAL MANAGEMENT

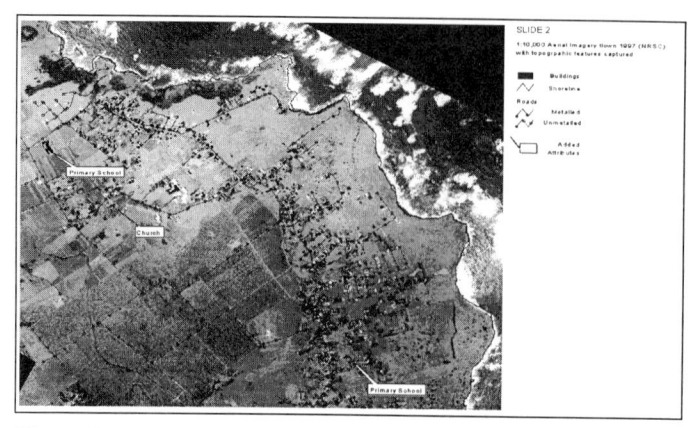

Figure 2 – Aerial Photography Overlain by Digital Linework

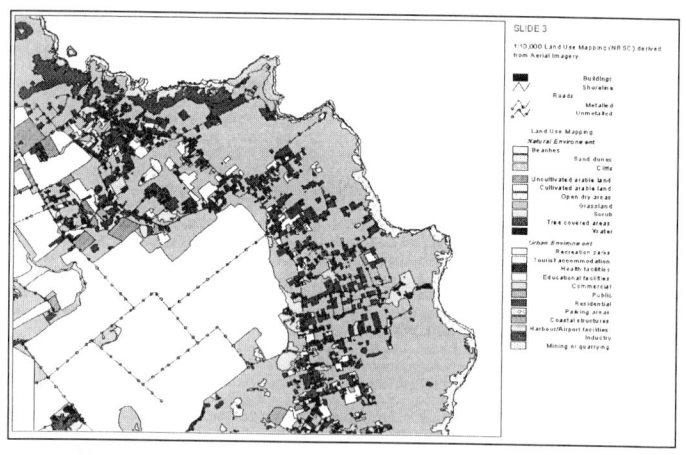

Figure 3 – Example of Land Use Mapping

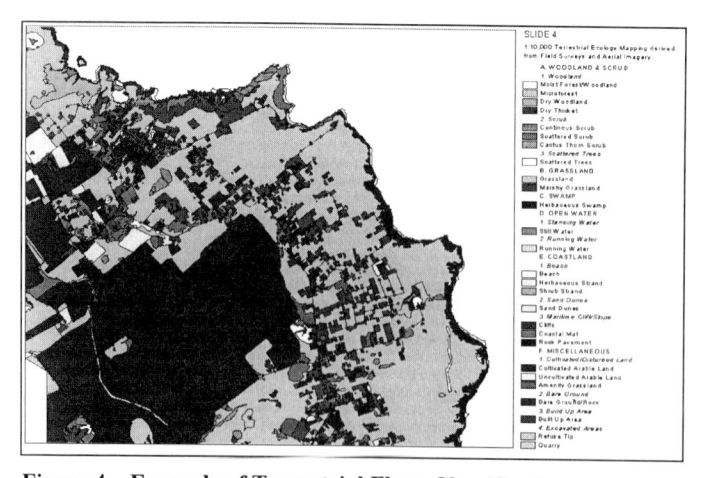

Figure 4 – Example of Terrestrial Flora Classification

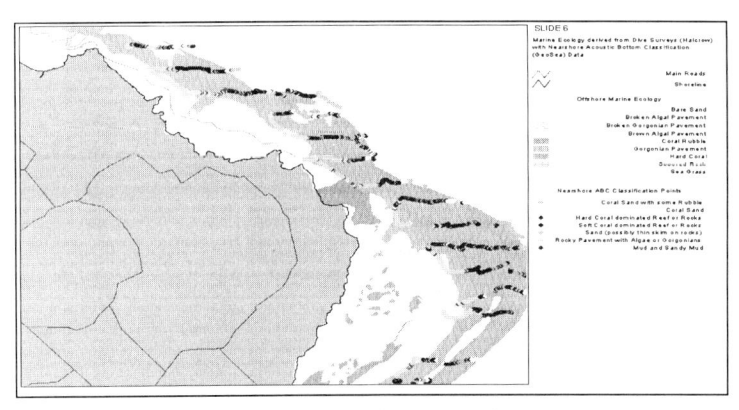

Figure 5 – Example of Marine Habitats Mapping

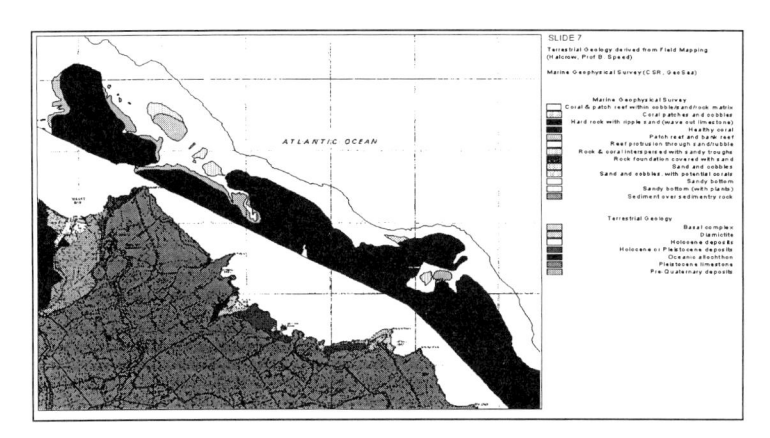

Figure 6 – Example of Mapping of Offshore Geophysical and Onshore Geology Survey Data

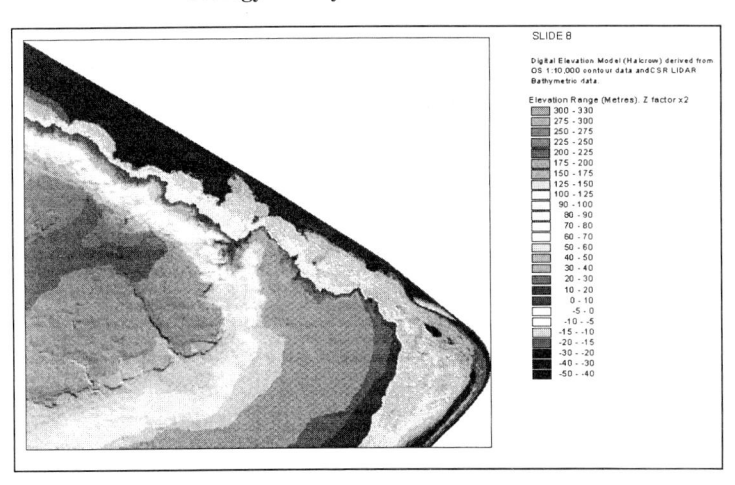

Figure 7 – Example of Merged Topography and Bathymetry Data Sets

The above field surveys have been supplemented by:

- Desk studies based upon available research of the onshore fauna (Halcrow, 1998l) and offshore fishery of the Plan area – in the former case a local specialist who is a recognised authority on turtles was utilised, whilst the fisheries assessment, brought together by an international specialist, was based upon the government's own statistics.

- Baseline economic studies – bringing together existing data on land values and planning applications to assess development and economic pressures (Halcrow, 1999b).

- Wide ranging assessments of environmental value (Halcrow, 1999b) – analyses were undertaken to assess

    - The value to Barbadian families of recreational outings to the Caribbean and Atlantic Coasts
    - The non-use (existence) value of the Atlantic Coast to Barbadians ie the inspirational value to nationals of this relatively remote coastline in its undisturbed state
    - The effect of the environmental quality of the coastal zone to stimulate repeat visits by tourists (both cruise ship and stay-over visitors)
    - The value of the coastal amenity and environmental quality to landowners on both the Caribbean and Atlantic Coasts
    - The value of the Caribbean Coast to short term Barbadian beach users

- Computer modelling of the coastal environment to investigate wave, current and sediment movement characteristics (Halcrow, 1998j) - the model investigations included

    - Hindcasting of waves from wind data
    - Regional wave transformation modelling
    - Detailed modelling of extreme waves for the entire study area
    - Modelling of wave induced currents using the results of the transformation model
    - At selected sites, the modelling of wave transformation, water level set-up, cross-shore transport and beach response during storm events
    - Estimation of potential alongshore transport rates using the results of the regional wave transformation modelling
    - Modelling of nearshore currents using the offshore measurements from OSCR
    - Grid based modelling of nearshore sediment transport using the results of the current and wave modelling.

PLAN DEVELOPMENT AND PRESENTATION

Alongside the range of field programmes and studies undertaken for the Atlantic Coast, the project was able to develop a vision for the future for the entire coastline of Barbados and establish strategic objectives and related policies through which this future for the coast might be achieved. Armed also with knowledge about those issues (Halcrow, 1998m) which consultees felt the Atlantic Coast Plan was obliged to address, the project was able to establish the themes and objectives most appropriate for each of the five sub-areas into which the project's coastline had been divided. An appraisal of these issues led the team to focus upon twelve topic areas and to establish general management guidance appropriate to each.

For each area the Plan provides an assessment of its current status, gives reasons for any concern and outlines a range of actions through which future management and monitoring of the coastal zone might more effectively be achieved. In every case the organisation needing to lead each component, a timescale for the completion of each action and suggested indicators through which achievement might be recognised have been presented.

Having established such general guidance, the Atlantic Coast Plan picks up on specific actions insofar as they relate to each of the sub-areas. Again the agencies involved in the implementation of each action and a timeframe have been provided.

Implementation of both ICM Plans is recognised as being a long term and continuous process. Some elements are seen as requiring immediate attention and are readily incorporated into the day-to-day operations of the responsible agencies; in others there is scope for building onto existing arrangements. New procedures and practices will have to be introduced to address the considerable extent and variety of topics. The Plans discuss those requirements considered to be essential for their successful implementation. They focus upon:

- the role of the community and how they should be engaged
- the role of environmental education, formally through schools, via adult education, and through tertiary level institutions
- structured public information and awareness programmes
- future information and data management needs
- a suggested role for indicators
- the role of the CZMU in implementing both plans
- the means through which the costs of Plan implementation might in the future be recovered.

THE WAY FORWARD

At the end of the previous South and West Coast commission, the CZMU were confronted with the implementation of a Plan (then still a draft) covering the island's Caribbean Coast. Integral with this however was the legislation needed to provide the legal foundation to the implementation of ICM in Barbados. Though various forms of the legislation have been put forward since that time, finalisation has proved to be difficult. A similar situation confronted the CZMU with the issue of the Atlantic Coast Plan. Amongst other things this has meant that the Plan covered by this paper continues to be considered as a draft. Considerable strides have however been made since that time, most notably with the passing in December 1999 of the Coastal Zone Management Act. The Act requires that the CZMU present both draft plans and an order delimiting a Coastal Zone Management Area to a public enquiry within 6 months of it being passed. Subject to the outcome of this and any revisions that are then determined to be necessary, both Plans will come into force following Ministerial approval. To assist this process, the Government saw the need for both Plans to be presented in a similar format. It is to be this reformatted version of the South and West Coast Plan that will be tabled with the Atlantic Coast Plan during the public enquiry.

Whilst in many ways implementation of both Plans is to be achieved by existing resources, our commission was able to identify deficiencies in knowledge, particularly for the environments of the South and West Coasts, that would benefit from early attention. The CZMU are in the process of commissioning new survey work and research to resolve some of these. Other actions requiring resolution appear likely to have to await the implementation of an all-encompassing (Phase II) investment programme. As currently presented by the team, the programme offers:

- a broad mix of large and small engineered investments
- measures to enhance public access to the coastline
- to help facilitate the engagement of the public in programmes that widen their understanding of and commitment to the coastal environment
- measures that protect the special status areas identified under both Plans
- a more general form of support and research that Government bodies may require to ensure the successful implementation of the Plans.

In taking this forward it remains for the CZMU to

- review these investments and modify the programme to meet any final budgetary constraints
- approach the relevant international lending institutions for determining a loan agreement to undertake the works
- ensure the availability of suitably trained staff to meet the mandated responsibilities of the CZMU and other arms of Government under the Act (Halcrow, 1998n)
- contract the resources needed to instigate and successfully complete the Phase II programme
- ensure that all other Plan obligations are met within the timescales indicated
- ensure through their regular review the continued appropriateness of the measures identified for action under both Plans

A great deal of work has already been carried out to promote and implement an effective programme of ICM in Barbados. The benefits are already evident, for example through the maintenance and protection of beaches, safeguarding the amenity value of coasts, and helping to conserve coastal landscapes and wildlife. Barbados has with considerable justification been able to promote itself as a regional leader in the development and implementation of ICM. It has thus been a great pleasure working with all levels of the CZMU to help them to make real progress towards the realisation of their vision for the coast.

REFERENCES

DELCAN (1995a) The Integrated Management Plan for the South and West Coasts of Barbados. Government of Barbados, Ministry of Tourism, International Transport and the Environment, Coastal Conservation Project Unit. Feasibility Studies on Coastal Conservation.

DELCAN (1995b) Final Investment Phase Report. Government of Barbados, Ministry of Tourism, International Transport and the Environment, Coastal Conservation Project Unit. Feasibility Studies on Coastal Conservation.

Halcrow (1997a) Lidar Survey Report. Government of Barbados, Ministry of Health and the Environment, Coastal Zone Management Unit. Barbados Coastal Conservation programme – Phase 1.

Halcrow (1997b) Socio- Economic Profile of Barbados Coastal Communities ((Maycocks Bay to South Point). Government of Barbados, Ministry of Health and the Environment, Coastal Zone Management Unit. Barbados Coastal Conservation programme –Phase 1.

Halcrow (1998a) Integrated Coastal Management Plan for the South-East, East and North West Coasts of Barbados. Government of Barbados, Ministry of Health and the Environment, Coastal Zone Management Unit. Barbados Coastal Conservation programme –Phase 1.

Halcrow (1998b) Integrated Coastal Management - The Barbados Policy Framework. Government of Barbados, Ministry of Health and the Environment, Coastal Zone Management Unit. Barbados Coastal Conservation programme –Phase 1.

Halcrow (1998c) Onshore and Offshore Geology and Geomorphology of Eastern Barbados. Government of Barbados, Ministry of Health and the Environment, Coastal Zone Management Unit. Barbados Coastal Conservation programme–Phase 1.
Annex 1 – Onshore Geology and Geomorphology
Annex 2 – Marine Bathymetry and Geophysical Survey
Annex 3 –Sediment Trends and Acoustic Bottom Classification Fieldwork Operations Report
Annex 4 – Sediment Trends Report

Halcrow (1998d) Land Use Planning Report. Government of Barbados, Ministry of Health and the Environment, Coastal Zone Management Unit. Barbados Coastal Conservation Programme - Phase 1.

Halcrow (1998e) Terrestrial Ecology Report. Government of Barbados, Ministry of Health and the Environment, Coastal Zone Management Unit. Barbados Coastal Conservation Programme - Phase 1.

Halcrow (1998f) East Coast Structures Inventory and Assessment. Government of Barbados, Ministry of Health and the Environment, Coastal Zone Management Unit. Barbados Coastal Conservation Programme - Phase 1.

Halcrow (1998g) Marine Ecology Report. Government of Barbados, Ministry of Health and the Environment, Coastal Zone Management Unit. Barbados Coastal Conservation Programme - Phase 1.
Annex 1 - Use and development and management options for coastal fisheries and production systems of the eastern coastal sector of Barbados.
Annex 2 - Acoustic Bottom Classification.

Halcrow (1998h) Hydrogeology, Terrestrial and Marine Water Quality. Government of Barbados, Ministry of Health and the Environment, Coastal Zone Management Unit. Barbados Coastal Conservation Programme - Phase 1.

Halcrow (1998i) Government Analytical Services: Institutional Strengthening Report. Government of Barbados, Ministry of Health and the Environment, Coastal Zone Management Unit. Barbados Coastal Conservation Programme – Phase 1.

Halcrow (1998j) Oceanography Report. Government of Barbados, Ministry of Health and the Environment, Coastal Zone Management Unit. Barbados Coastal Conservation Programme - Phase 1.

Halcrow (1998k) Integrated Coastal Management Plan for the Atlantic Coast - Consultation Summary. Government of Barbados, Ministry of Health and the Environment, Coastal Zone Management Unit. Barbados Coastal Conservation Programme - Phase 1.

Halcrow (1998l) Terrestrial Ecology Report (Fauna). Government of Barbados, Ministry of Health and the Environment, Coastal Zone Management Unit. Barbados Coastal Conservation Programme - Phase 1.

Halcrow (1998m) Issues Analysis. Draft report. Government of Barbados, Ministry of Health and the Environment, Coastal Zone Management Unit. Barbados Coastal Conservation Programme - Phase 1.

Halcrow (1998n) Capability and Capacity of the Coastal Zone Management Unit. Government of Barbados, Ministry of Health and the Environment, Coastal Zone Management Unit. Barbados Coastal Conservation Programme - Phase 1.

Halcrow (1999a) Legal Update and Strengthening Report for Integrated Coastal Zone Management Planning. Government of Barbados, Ministry of Health and the Environment, Coastal Zone Management Unit. Barbados Coastal Conservation Programme - Phase 1.

Halcrow (1999b) Natural Resources and Socio-Economic Studies Report. Government of Barbados, Ministry of Health and the Environment, Coastal Zone Management Unit. Barbados Coastal Conservation Programme - Phase 1.
Annex 1 - The value and use of Barbados's east coast sites for recreation
Annex 2 - Measurement of east coastal zone environmental quality through land values
Annex 3 - A model for indicative land use change in the eastern coastal zone.

Halcrow (1999c) Littoral Processes Report. Government of Barbados, Ministry of Health and the Environment, Coastal Zone Management Unit. Barbados Coastal Conservation Programme - Phase 1.

Economic Evaluation of Coastal Management in Japan

MR Y. IMAMURA, MPhil, MEng, BEng, Deputy Director, Ministry of Construction (MOC), Tokyo, Japan

INTRODUCTION

Although construction of the infrastructure has contributed to the progress of our society, an increase in efficiency and transparency is needed in the planning of a project because of a reduction in the working population, financial limitations and a more appropriate allocation of the budget in Japan[1]. Economic evaluation is regarded as very useful tool to solve the problem. It was in this context that the first economic appraisal scheme for coastal management in Japan was developed.

In this paper, firstly, the context of the issue is explored. Secondly, the process of the development is described. Thirdly, the content of the scheme is explained. Fourthly, the implementation of the scheme is set out. Fifthly, problems and current challenges are discussed. Finally, our current results and future steps are explored.

CONTEXT

Many urban areas in Japan, especially large cities, have developed on the coastal lowland because most inland areas are mountainous and the fluvial lowland is suitable for the cultivation of rice which is the staple diet of the Japanese. It is apparent that sea defence is essential as a governmental work because coastal disasters have caused enormous damage to our country. For example, 27,123 people died in the Sanriku Tsunami Disaster in 1896[2]. Even after World War II, 4,697 people were killed and more than 150,000 houses were destroyed by the Ise Bay High Tide in 1959[2]. Therefore, not only coastal engineers but also the public did not think an economic appraisal was necessary in the planning of coastal protection. Moreover, the annual budget for coastal protection is less than 10% of that for river management (about £10,000 million), including fluvial flood defence and water resource development, and the number of coastal engineers is much less than that of river engineers. Thus, the development of an economic appraisal for coastal management lagged behind, although river engineers published guidance for an economic appraisal of flood defence in 1970[3].

Coastal Management: Integrating science, engineering and management, Thomas Telford, London, 2000

There are two reasons why we must develop an economic appraisal method for coastal management. One is that people require efficacy and transparency in public works and economic analysis is the effective tool for it. In particular, an economic appraisal is essential in the assessment of projects, with regard to both planning and reviewing; and planning and reviewing are becoming important. The other is an increase in conservation works, where its priority is not as clear as with sea defence although concern for the environment is growing.

PROCESS OF DEVELOPMENT OF ECONOMIC APPRAISAL SCHEME

The principal aim of coastal management in Japan is sea defence and this defence has the greatest similarity to flood defence. Therefore, the knowledge and experience of the economic appraisal of fluvial flooding since 1970, was utilised and the first version for coastal management was developed in 1996. The technical advisory committee, which consists of five authorities in economics, planning, environment and coastal engineering, was organised in September 1997 by the Ministry of Construction (MOC) with the Ministry of Transport, and the Ministry of Agriculture, Forestry and Fisheries to give us advice[4]. Until June 1998, the committee met four times and made recommendations to us for our draft version. Then, 'Guidance of Economic Appraisal of Coastal Management 1998'[5] was authorised in October 1998 after the examination of the recommendations and our official procedure.

ECONIMIC APPRAISAL SCHEME
Objectives

The objectives of the scheme are expressed as:
- For transparency, so that the method and result is understandable to the public.
- For the appropriate allocation of a limited budget for coastal management, the effect of projects is evaluated objectively.
- For practical use, the scheme is easy to handle by all coastal engineers nation-wide.

Basic approach and scope

The scheme is looking at economic appraisal not of nation-wide coastal management but of each project. Cost benefit analysis (CBA) is used to assess the economic feasibility of a project. In the analysis, costs and benefits are counted as at each year and converted into net present value (NPV). They are calculated both with and without a project. The balance between both cases is defined as the cost and benefit of the project. Then, the cost and the benefit are compared.

Term of a project appraised

The term of an appraisal starts with the beginning of a project and ends with the finishing of the effect of a project. The effect of a project is assumed to last for fifty years after completion of the whole or part of a project.

Benefit
In the scheme, three kinds of benefit are measured.

The first is the benefit by a reduction of inundation damage. Many casualties and a huge loss of property have been caused by high tides and tsunamis in our country. Due to sea defence, such as the construction of sea walls and detached breakwaters, inundation losses are averted. There are two kinds of damage by inundation as follows:
- Damage to physical stock including agricultural stock.
- Intangible costs, such as mental shock and mental exhaustion.

Although, according to recent research[6)7)], it is said that intangible losses are important components in the estimation of the impact, the method of quantifying them in monetary term is still under development. Therefore, the intangible damage is not counted in this scheme.

The second benefit is the prevention of coastal erosion. According to a study by the Public Works Research Institute, MOC, 160 ha of coastal beach per year were lost by erosion in Japan from 1978 to 1991[8)]. It is said that the speed of the erosion accelerated after the Second World War and the erosion has not stopped yet. There are three kinds of impact caused by coastal erosion.
- Damage to physical stock including agricultural stock
- Intangible costs, such as mental shock and mental exhaustion
- Loss of beaches which are essential to proper coastal management

The second and third impacts cannot be quantified in economic terms at the moment. Accordingly, diminution of physical damages is calculated in the scheme.

The final benefit is conservation of nature and the improvement of coastal recreation. In preserving the beautiful landscape and nature on coasts, people can enjoy nature both directly and indirectly.

The total benefit is valued by summing up the three above benefits avoiding duplication.

Cost
Costs consist of two categories. One is the cost of construction, including building of breakwaters and nourishing beaches. This cost is classified as 'capital' or 'initial' cost. The other is the maintenance cost. Although maintenance is an essential element for proper coastal management, it has often been miscalculated or even ignored. However, it is vital for the

economic assessment of alternatives. In the case of the shore protection project managed by MOC in Kaike Coast, which is located in the western part of Japan, sand recycling and construction of detached breakwaters were compared. Without accurate calculation of the maintenance cost, it had been believed that construction of detached breakwaters is more economical than sand recycling because of continuous running costs for sand recycling while disregarding discount. The computation by this new method showed that the total cost of sand recycling for one hundred years was less than a half of that of construction of detached breakwaters and sand recycling was selected after all[9]. In estimation of the maintenance cost, summing it up each year is the best way. If it is difficult to calculate it throughout the term of a project, 0.5% of capital cost per year can be used. The ratio, 0.5%, is quoted from 'Guidance of Economic Appraisal for Flood Defence 1970[3].' If land acquisition is needed in a project though it is a very rare case in Japan, the cost should be included. As for tax, value-added tax must be deducted from the initial and maintenance cost.

Sequence of appraisal
The sequence of appraisal is shown in Figure 1.
- The premise for economic appraisal, such as the natural and social conditions of a coast and the related strategy and policy on a project, is identified.
- The scope of CBA is determined.
- Three kinds of benefit are calculated and the total benefit is measured in each year.
- The total costs, including the construction and maintenance cost, are estimated in each year.
- The benefit and cost in each year are converted to net present value (NPV), and the total benefit and cost through a project term are calculated.
- The total cost and total benefit are compared.

Benefit by reduction of inundation damage
Coastal inundation is caused by high tides, waves, or tsunamis. The basic approach to them in economic appraisal is common. In this paper, the case of high tides and waves is described.

Firstly, the probability of high tides and waves is estimated. Then, a sequential sea level model is made at each probability. Generally, five cases of probability are selected from one in ten years to fifty years. The duration time of a sequential model is twelve hours in an enclosed sea and twenty-four hours in an open sea. If there is a model in a sea defence programme, the model should be used. Next, a volume of overflow is calculated. A database of land elevation in a potential inundation area is also created. Using the results of them, the area and depth of inundation is computed at each probability. Together with it, the property in the inundation area is valued. The property is classified into six categories.
- Residential property: houses and buildings, and household goods.

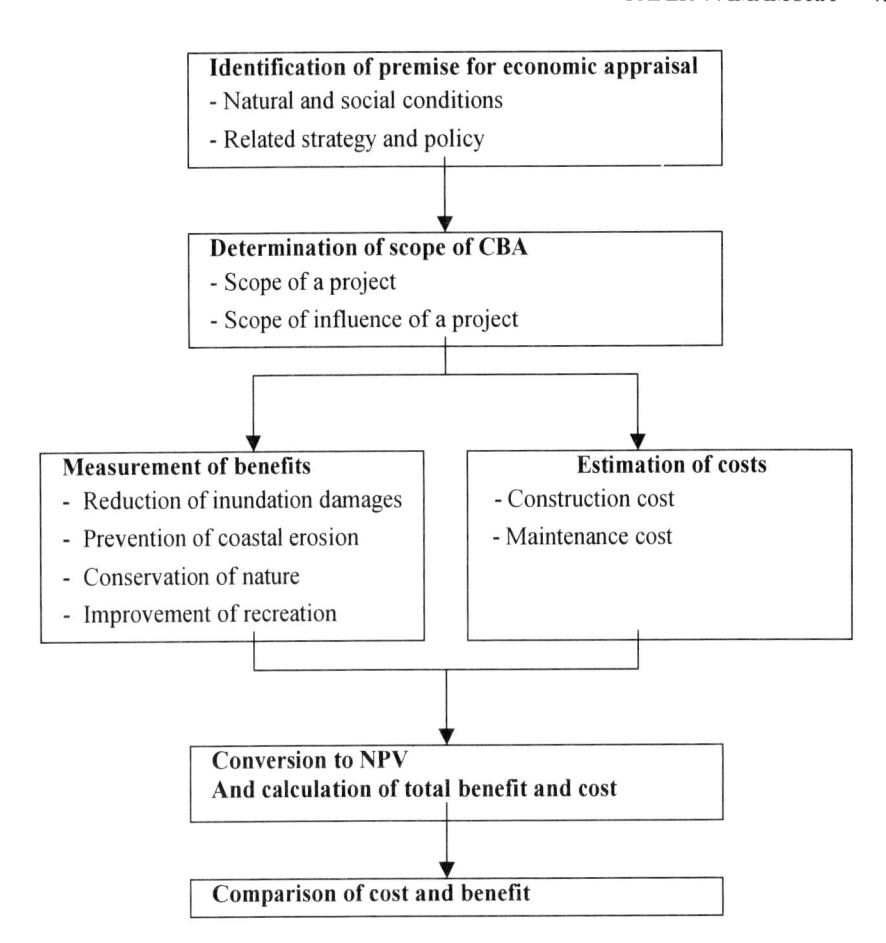

```
┌─────────────────────────────────────────────┐
│ Identification of premise for economic appraisal │
│ - Natural and social conditions              │
│ - Related strategy and policy                │
└─────────────────────────────────────────────┘
                     │
                     ▼
┌─────────────────────────────────────────────┐
│ Determination of scope of CBA                │
│ - Scope of a project                         │
│ - Scope of influence of a project            │
└─────────────────────────────────────────────┘
             │              │
             ▼              ▼
┌──────────────────────────┐ ┌──────────────────────────┐
│ Measurement of benefits   │ │ Estimation of costs       │
│ - Reduction of inundation │ │ - Construction cost       │
│   damages                 │ │ - Maintenance cost        │
│ - Prevention of coastal   │ │                           │
│   erosion                 │ │                           │
│ - Conservation of nature  │ │                           │
│ - Improvement of recreation│ │                          │
└──────────────────────────┘ └──────────────────────────┘
                     │
                     ▼
┌─────────────────────────────────────────────┐
│ Conversion to NPV                            │
│ And calculation of total benefit and cost    │
└─────────────────────────────────────────────┘
                     │
                     ▼
┌─────────────────────────────────────────────┐
│ Comparison of cost and benefit               │
└─────────────────────────────────────────────┘
```

Figure 1. Sequence of appraisal

- Commercial and industrial property: redeemable goods and stocks.
- Agricultural and fishery property: redeemable goods and stocks.
- Agricultural production: losses of agricultural production for five years caused by sea water inundation
- Facilities constructed by public works, such as roads, bridges and parks.
- Facilities constructed by public utility works, such as electric, gas and water supply services

Then, damages are estimated using damage-depth ratio for residential property, commercial and industrial property, and agricultural and fishery property. Loss of agricultural production is counted using damage ratio. The impact on facilities constructed by public works and public utility works is measured from total damage of general property* using damage ratio between general property and facilities constructed by public works and public utility works. The ratio is calculated using the past twenty years data of flooding. The total of these damages is the estimated annual loss.

General property *: residential property, commercial and industrial property, and agricultural and fishery property

Benefit by prevention of coastal erosion
The effect of prevention measures is equivalent to the loss that would have occurred by coastal erosion without a measure.

Firstly, the annual speed of erosion is estimated by referring to the change of shoreline in the past. Then, the land area erosion is inferred. Next, the land area is divided into five categories; residential area including commercial and industrial area, agricultural area, forest area, road and beach. The land price of each category is assessed using the transaction price, published price or price for property tax around the area. Property in the land area is valued applying the same method as in the calculation of benefit by reduction of inundation damage. Consequently, the annual total loss of land and property is computed.

Value of conservation of nature and improvement of coastal recreation
The values of conservation of nature and improvement of coastal recreation can be identified in various ways. One example of classification is as follows[5]:
- Existence value to conserve natural ecosystem
- Existence value to preserve natural landscape
- Use value in improving coastal activities, such as recreation and sports
- Existence value to gain amenity for the public

No specific way to appraise the values is described in the guidance, but the Contingent Value Method (CVM) is recommended and is required of accumulation of research on CVM to improve the reliability of the method.

Conversion to Net Present Value (NPV)

First of all, the base year must be determined. In the guidance, the starting year of a project, or the year to begin to occur benefit or cost is suggested as the base year. Next, the influence of past inflation is removed from benefits and costs in each year using a price index, such as the index for coastal management in the latest index note for economic appraisal for flood defence[10]. Then, the annual benefit and cost are converted to NPV using a 4.0% discount rate. The rate of 4.0% is proposed by the task force in MOC for all public works referring past interest rates and other economic data[11].

Benefit (NPV) $= Bt / (1+i)^t$
Cost (NPV) $= Ct / (1+i)^t$
Bt: annual benefit in t year
Ct: annual cost in t year
i: discount rate

Comparison of total benefit and cost

Three methods shown below are widely used in the Cost-Benefit Analysis.

- Net Present Value Method

$$NPV = \sum_{t=1}^{n} Bt / (1+i)^t - \sum_{t=1}^{n} Ct / (1+i)^t$$

- Cost Benefit Ratio Method

$$CBR = \sum_{t=1}^{n} Bt / (1+i)^t / \sum_{t=1}^{n} Ct / (1+i)^t$$

- Internal Ratio of Return Method

$$\sum_{t=1}^{n} (Bt - Ct) / (1 + i_0)^t = 0$$

In the guidance, the NPV method and CBR method are proposed to judge the economic feasibility of a coastal management project.

IMPELEMENTATION OF THE SCHEME

In Japan, coastal management projects are divided into three types:
- projects managed by the national government
- projects managed by local government and subsidised by the national government
- projects managed by local government with no subsidy

MOC began to utilise the scheme to make the economic feasibility of projects, including projects subsidised by it, clear. Together with it, the review programme of public works including an economic assessment was launched by the national government in 1998[12].

Therefore, eleven coastal projects by MOC were reviewed and the scheme was used in the review. To appraise the economic feasibility of all projects of the second type which start in the 1999 fiscal year, the scheme was applied as well. In the review of the second type of project, the scheme was implemented. The results of the appraisals were available through press releases. The public perceived that the scheme helped to increase the transparency in the appraisal of projects and no strong criticism for the scheme has occurred at the moment.

CURRENT CHALLENGES FOR THE NEXT STEP

Although cost benefit analysis (CBA) is a powerful tool to assess the economic feasibility of a project objectively, it has some problems as shown below [4]:

- The method is useful for comparing the alternatives of a project and the economic efficiency among similar projects: same type and scale. However, it is not suitable to weigh against different types of projects. For example, in the case that the benefit of a shore protection work is evaluated by the surrogate markets method and that of conservation is counted by CVM, their sensitivity and reliability are not equivalent. Therefore, it is not reasonable to decide the priority between them using the output of their CBA.
- The technique is still developing and the benefits capable of being calculated are limited. For example, we have not found a satisfactory way to evaluate intangible effects. Saving human lives is one of the most important aims of coastal management. However, it is said that the value of a human life as a contingency dealt statistically and that in a specific context are heterogeneous issues[13].

Among our challenges for solving these problems, research[14,15] on intangible effects and CVM are explored in this paper.

Intangible effects due to coastal disasters were investigated on the basis of a questionnaire distributed among coastal residents. Sampled people who have not experienced coastal disasters tended to feel more fear and uneasiness about coastal disasters. Those who have experienced coastal disasters generally feel fear and uneasiness at the occurrence of the disaster and during evacuation. Discomfort gradually tends to dominate in the recovery stage. Further analysis revealed that the intangible effects due to coastal disasters were classified into three patterns depending on the amount of tangible damage. A questionnaire sheet was presented, which will be used to estimate the intangible damage in monetary term based on the Contingent Valuation Method.

The benefit of beach preservation at Niigata coast, which is located on the Japan Sea in the central Japan, was estimated by the Contingent Valuation Method. Estimation was performed from the environmental aspect on the beach recovery efforts of Niigata coast by coastal management projects. The result was correlated with various properties, such as age, sex, annual income and the distance of the residence from the coast. It was found that median value of 1,591 yen/year could be afforded for the maintenance of the present beach environment for the use of recreational purposes. High correlation was found between the interest in beach preservation and the frequency of visits to the beach.

CONCLUSION

The first economic appraisal scheme of coastal management in Japan was developed in October 1998. It utilises cost benefit analysis and the main component of the benefit is the reduction in damage by inundation and erosion. The scheme is implemented to assess the economic feasibility of coastal management projects nation-wide. The results of them are open to the public and we are receiving a good response from people.

By using the economic appraisal scheme, the planning of a coastal management project has been much advanced and it is the first big step. Nevertheless, there remain some problems described in 'Current challenges for the next steps.' Our mission is expressed as:
- To inform the public of the imperfection of the scheme.
- To make every effort to reduce the obstacles.
- To assess the feasibility of a project objectively utilising the fruits of the efforts.

Finally,
- To open the results to people to increase public understanding.

We are continuing our challenge for the next step[16]. In 2000, it is planned to launch a new version of the guidance and the results of the current challenge are to be used in it.

REFERENCES

1) Construction Policy Research Centre, MOC, 1997. *Benefit evaluation of infrastructure development*, Policy Research Centre Note 14.
2) Seacoast Division, MOC, 1998. *Coastal Management in Japan*, p8.
3) River Planning Division, MOC, 1970. *Guidance of Economic Appraisal for Flood Defence 1970*.
4) Imamura, Y, Orito, M, 1999. *Cost-benefit analysis on coastal management*, Seacoast 38-2, pp80-84.
5) Ministry of Construction (MOC), Ministry of Agriculture, Forestry and Fishery, and Ministry of Transport, 1998. *Guidance of Economic Appraisal of Coastal Management 1998*.
6) Kuriki, M, Imamura, Y, Kobayashi, H, 1996. *Economic Evaluation of Intangible*

Effects of Flooding, Natural Disaster Science 15-13, pp231-240.
7) Imamura, Y, et al., 1994. *Intangible Effects of Flooding*.
8) Public Work Research Institute, MOC, 1992. *Investigation on coastal erosion*.
9) Itou, H, 1999. *Economic evaluation of sand recycling in Kaike Coast*, Seacoast 38-2, pp90-93.
10) River Planning Division, MOC, 1998. *Index note for economic appraisal for flood defence*.
11) MOC, 1998. *Guideline for Appraisal of Public Works*.
12) Imamura, Y, Okayasu, T, 1999. *Newly launched programme of reviewing and planning of coastal management projects*, Seacoast 38-2, pp85-89.
13) Kobayashi, K, 1998, *Benefit evaluation of safety and environment*
14) Imamura, Y, Kawase, H, Itou, Y, Sato, S, Kasai, M, Morota, I, Hirano, G, 1999. *Investigation of intangible damage about coastal disaster*, Civil Engineering in the Ocean 15.
15) Imamura, Y, Sato, S, Kasai, M, Saito, A, Hara, H, Hirano, G, 1999. *Investigation of intangible damage about coastal disaster*, Civil Engineering in the Ocean 15.
16) Imamura, Y, et al., 1998. *Evaluation of effects of coastal management*, Technical Research Conference (MOC) 52, pp1-1. -1-22.

All of the above references are written in Japanese and some of their titles are translated into English by the author if they do not have English titles.

Can Land Consolidation Schemes used in Denmark provide a mechanism to facilitate managed realignment in the UK?

DR. D.E.JOHNSON
Maritime Faculty, Southampton Institute, Southampton, UK

ABSTRACT
The need for and benefits of managed retreat/realignment along the rural coastline in areas of the UK currently suffering from "coastal squeeze" have now been debated for almost a decade. The option has been incorporated within shoreline management plans and a series of ongoing field experiments are yielding encouraging results, both in terms of ecological knowledge and engineering competence. However, a significant barrier to achieving the larger scale projects, which coastal defence planners envisage will be required in the future, is the Government's unwillingness to compensate private landowners, whose property will be lost to the sea, in the event of adopting either a "do nothing" or "managed retreat" strategy.

This paper reviews the effectiveness of a Danish tool for nature restoration which may provide a solution. In 1989 the Danish Nature Management Act enabled land consolidation schemes, or reallotment of land, which had previously only been used for agricultural purposes, to be used to integrate both environmental and agricultural objectives. Case studies of coastal schemes illustrate how a participative approach has been used to achieve broad-based local support without the need for compensation.

INTRODUCTION
Coastal planners and managers responsible for low-lying regions of the UK will increasingly need to make provision for accelerated relative sea-level rise. The extent of this impact, and thus areas threatened by inundation, has been variously predicted (Tooley, 1989; Carter, 1989; Boorman *et al.*, 1989; and Shennan, 1993) and is the subject of on-going research. However, for South-east Britain, affected by a combination of isostatic and eustatic change, the government (DoE, 1996) has predicted an annual rise of approximately 5cm/decade and a rise of +37cm by the 2050's. "Coastal squeeze", a combination of development or intensive agriculture up to and into the intertidal zone and landward movement of high and/or low water marks squeezing out intertidal habitat, places the range of environmental, social and economic benefits associated with intertidal wetland areas at risk. Specifically these include nature conservation, coastal defence, water quality, commercial fisheries, recreation, landscape quality and educational values. Pye and French (1993) estimated that by 2010 the UK could lose a further 4% of intertidal flats (8,000 - 10,000 ha) and 8% of saltmarshes.

The recognition that benefits associated with healthy functioning of intertidal wetland ecosystems are irreplaceable has prompted conservation efforts and the development of a range of potential solutions. These can be categorized in terms of a preservation - restoration continuum which is summarized in Figure 1. To counter relative sea-level rise, restoration

50   COASTAL MANAGEMENT

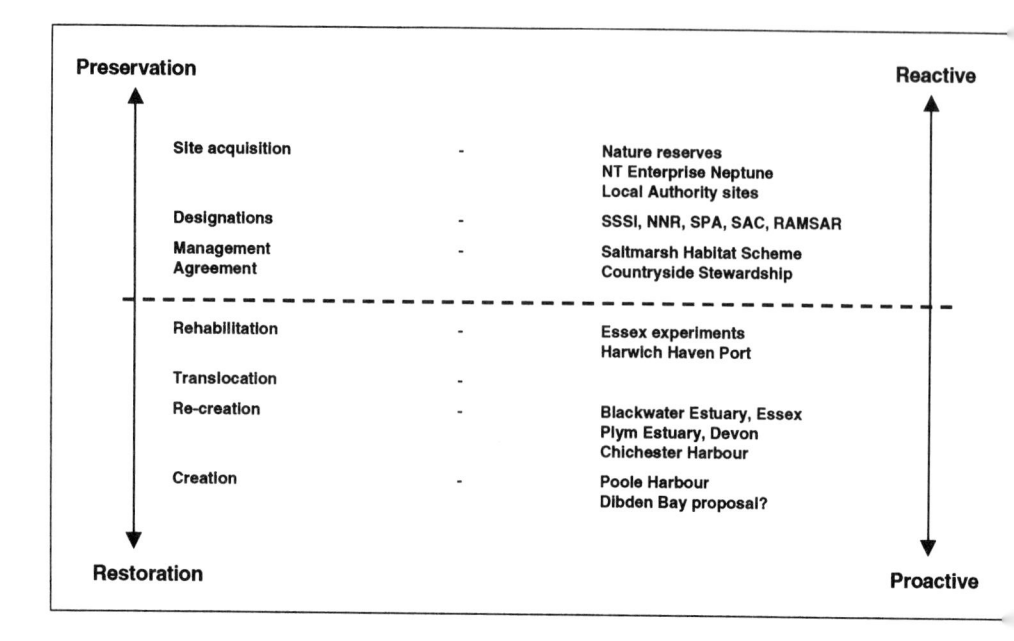

Figure 1: UK Intertidal Wetland Conservation Continuum

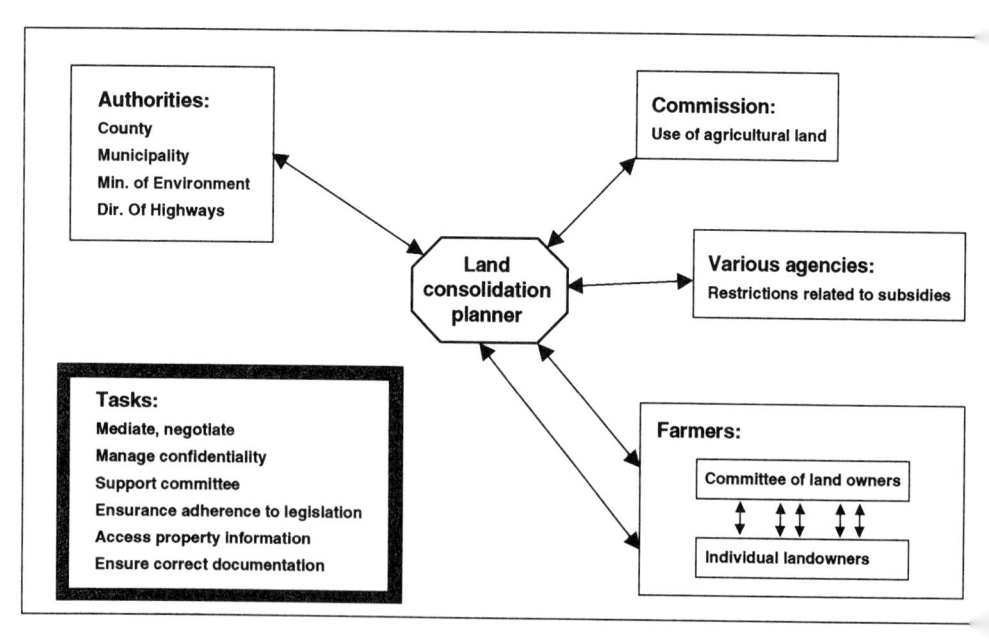

Figure 2: Role of the land Consolidation Planner (source: Danish Government)

techniques are required. These have been classified as surface, seaward or landward techniques (NRA, 1995). Surface and seaward techniques usually apply to habitat rehabilitation projects, whereas landward techniques generally apply to habitat re-creation projects. Surface techniques aim to restore the health of degraded intertidal ecosystems, without necessarily extending their area. Seaward techniques aim to extend intertidal habitat by promoting accretion, either by decreasing water velocity over the marsh or by increasing the sediment load that the water carries. Landward techniques aim to convert previously enclosed land to intertidal (MAFF, 1993). On this basis NRA (1996) identified four main strategies for maintaining and enhancing intertidal habitat namely: increasing vertical accretion; combating lateral erosion; improving vegetation cover and vigour; and "managed retreat" (also termed managed realignment or set-back).

Pethick (1996) maintained that any restoration programme for intertidal wetlands must allow for a redesign of estuaries to enable them to function properly. His argument is that the coast has more energy than we are able to control using present technologies, and that by initiating coastal squeeze man has "hijacked" and "canalised" estuaries for sectoral ends, thereby increasing tidal range and increasing risks to both habitat and coastal property assets. On this basis optimum locations can be identified for re-creation with the tidal prism of the estuary, estuary morphology, site history and surface elevation and gradient being considered as key criteria. For landward restoration projects, technical considerations must also include the balance of soft and hard engineering solutions, beneficial use of dredged material, sediment containment, performance criteria for restoration, optimal channel network design and the value of mathematical modelling (Carpenter and Pye, 1997).

Full-scale experimental research projects continue to inform current thinking. For example, set-back need not necessarily involve completely breaching the sea wall, if the wall is in good condition; instead a network of meandering creeks might be excavated inland, fed by sleeved sluices, as has been achieved on the Abbots Hall Estate in Essex (Dixon *et al.*, 1998).

ICZM ISSUES FACING INTERTIDAL WETLAND RE-CREATION IN THE UK

Clearly, whilst greater engineering certainty is desirable, there can be little doubt that intertidal wetland re-creation is technically possible. A fundamental weakness in the UK, however, is the absence of any national ecological restoration policy for intertidal wetlands. Shoreline Management Plans have confirmed options which are essentially both practicable (i.e. in the light of the physical environment, landward land use and existing coastal protection works) and cost-effective, but little attempt has been made to integrate biodiversity targets, partly because many have yet to be set. To address this a strategic approach has been expounded by Collins *et al.* (1997) which argues for estuary-based re-creation targets and integration of intertidal biodiversity targets with shoreline management planning.

Furthermore, whilst it is likely that investment by society in intertidal wetland restoration will not be successful without the requisite scientific competencies gained from an understanding of the biophysical requirements; at the same time appropriate restoration cannot proceed, especially beyond experimental pilot projects, without appropriate funding. Central government should pay for this work. The level of sectoral interest is too low and it is unrealistic to leave the funding to local authorities with other demands on their budgets. Additional core funding will need to be found by both DETR and MAFF. There is a real opportunity to achieve this, over and above existing incentives, as part of the present debate about redirecting Common Agricultural Policy subsidies to support conservation benefits instead of food production surpluses[1]. An element of local authority funding can also be argued for however, on the basis that the public benefit to communities within coastal areas should be funded from the public purse (i.e. through Council Tax). The local authority level is also the most important scale to take forward strategic plans and to commission and audit major works. Townend (1997) has proposed a 60/40 (central/local government) funding split as one way of prompting more innovative solutions.

Finally, entrenched sectoral positions are a major barrier to resolving where intertidal wetland restoration, and in particular managed realignment, should take place. Sorensen (1997) is of the opinion that:

> *"a shoreland exclusion or management law, or executive decree, appears to be the most cost-effective technique in the ICZM tool box".*

He identifies 25 nations with this type of legal mechanism. One of the elements emerging from the EU ICZM Demonstration Programme (EC; 1998) as a key to successful ICZM is also the need for mechanisms for compensation and controlled - but compulsory - acquisition of coastal property by authority. A priority must be for government to produce a mechanism acceptable to all stakeholders which both presents and enhances the existing intertidal wetland resource. A mixture of designation, mitigation, land purchase, compensation and social marketing has to be found. Its purpose would be to remove development value and development rights along the coastal strip on a selective basis.

A potential solution might be drawn from examining the Danish system of ecological restoration through land consolidation, which facilitates reallocation of land ownership for conservation purposes.

LAND CONSOLIDATION IN DENMARK

The principle of land consolidation is enshrined within the Danish agricultural system. Redistributive land reform has operated since the 1920's and its legal basis is the Land Consolidation Act. Another statutory instrument, the Act of Land Acquisition, facilitates acquisition of an initial "pool" of land for subsequent re-distribution in a land consolidation. The Act of Land Acquisition empowers the Minister to register a right of pre-emption on a property; to make this right of pre-emption effective when the property is sold; and to acquire land on the basis of "willing seller-willing buyer".

Land consolidation has been defined by the Danish Directorate for Development in Agriculture and Fisheries as:

> *"A procedure for simultaneous handling of a (sometimes large) number of sales and purchases of land with the effect that farmers exchange land and achieve a better location of their land. Fields are "moved" closer to the buildings and agricultural transport on roads is reduced."*

---

[1]MAFF's Habitat Scheme Saltmarsh Agreement currently operates throughout England on arable or permanent pasture grassland situated immediately adjacent to the coast. It offers a schedule of payments, over 20 year agreements, to create or extend saltmarsh.

The principal features of this land management tool are that:

i) land consolidation consists of a number of sales and purchases of land parcels which are made effective at the same moment in time;

ii) the farmers do not buy and sell from each other, sales and purchases go through the land consolidation and each individual owner only has one document irrespective of the number of transactions that person is involved with;

iii) a land consolidation planner (land surveyor or agronomist) negotiates with farmers and authorities (see Figure 2). The land consolidation planner is assisted by a committee of elected representatives of the land owners. He/she ensures that all rules concerning property transactions are adhered to and is responsible for registering the final result;

iv) all changes in the land consolidation scheme are made legally effective in a legal ruling by a land commission;

v) owners accept approximate indications of areas which means that cadastral surveys are only carried out when the changes are formally effective; and

vi) costs are paid by the state.

In practice the land consolidation planner works with two plans. Plan 1 represents the existing situation, Plan 2 the proposal. Plan 2 is used as the basis for negotiations. The final agreed version of Plan 2 must be approved by all relevant local authorities, including the local Commission for Agricultural Development (Ekner, 1995).

Until the late 1980's this procedure was restricted to amalgamation and structural adjustment of land holdings for agricultural purposes only. In 1989, in response to concerns regarding the environmental impact of intensive agriculture, the Nature Management Act extended the use of land consolidation to natural restoration. Reform of environmental and planning legislation during the 1990's has also fostered more detailed zoning of land uses coupled with grant payments (Sorenson, 1994).

Implementation of the Nature Management Act is based on a principle of voluntary participation. The rationale is that the public should gain maximum benefit in terms of environmental and recreational assets within a prescribed budget, and landowners should gain in terms of economic viability and operational convenience. Thus the system works on the basis of broad-based local support and cooperation between authorities rather than any form of compulsory purchase. The Act makes provision for expropriation and the recording of an option to purchase in the Land Register but these measures are regarded as extreme and have not been employed to date.

Nature restoration land consolidations are initiated by the National Forest and Nature Agency using an annual budget allocation. Independent assessments are made to ensure property is not acquired above local market prices and the land consolidation planner is a civil servant.

SELECTED CASE STUDIES
Three schemes, established on the Jutland coastline in the early 1990s, are described below to illustrate how a participative approach has been used to achieve broad-based local support.

Fjand Enge
Fjand Enge is a consolidation scheme of approximately 200 ha which involved 50 farmers in a project to re-establish traditional saltmarsh grazing.

This site, within the southern section of Nissum Fjord, is an internationally important nature conservation area. Its value relates to a mosaic of saltmarshes, brackish shallows and reedbeds which are dependant on traditional management. Attempts to cultivate the saltmarshes in the 1970s and fragmentation of ownership, preventing an effective grazing scheme, reduced the area's nature conservation value. The land consolidation scheme, initiated in 1990, with the intention of re-establishing high grade saltmarshes, enabled the formation of large enclosed areas for common grazing. Total set up costs were 2.8 million Danish kroner (DDK) for acquisition and land consolidation, plus a further 0.1 million DDK for fencing.

Legind Vejle
Legind Vejle is a consolidation scheme of 115 ha in which 17 properties have been reorganised to enable the re-establishment of an important coastal freshwater habitat.

This scheme is located on a former Fjord arm, cut off from the sea by geomorphological processes, which then developed as a brackish water lake. The lake reduced in size as a result of natural succession and was then drained for agricultural purposes from 1927 onwards. In the mid 1970s local residents petitioned the authorities to re-establish the Fjord arm as a nature reserve. At the time objection from a small number of landowners opposed to the proposal prevented its implementation. The 1989 Nature Management Act enabled the National Forest and Nature Agency to acquire a key property and effect the land consolidation. Voluntary redistribution of productive land, in exchange for low-lying land within the former Fjord arm, was carried out. The result is a shallow 27 ha lake with a range of associated freshwater wetland habitats. The total cost for this scheme was DKK 4.6 million comprising approximately DDK 1.2 million for preliminary surveys and project design; DDK 1.1 million for acquisition and land consolidation; DDK 2.2 million for construction works; and DDK 0.1 million for public relations.

Geddal Saltmarshes
Geddal saltmarshes is a consolidation scheme which involved 12 owners in a major intertidal wetland re-creation project. It was part funded by the EC ACE-Biotopes Programme.

In the 1950's, 140 ha of this site were reclaimed for intensive farming. A low "summer dyke", which had been constructed in the 1880s, was raised to prevent winter flooding and the salt meadows were drained. However, by the 1970s the raised dyke was proving difficult to maintain. Negotiations were progressed between land owners and the National Forest and Nature Agency to transfer the whole area to public ownership. In 1992, on the basis of a land consolidation agreement, the saltmarsh habitat was re-created by lowering the dyke along the seaward margin from 2-3m to 1.25m. Saltwater now inundates the area after high tides in winter. A new overflow replaced the drainage system and a secondary dyke was constructed inland to protect local property. Traditional management, cattle grazing and haymaking, have been resumed. Monthly monitoring records have recorded a significant increase in both breeding birds, including Avocet *(Recurvirostra avosetta)* and Ruff *(Philomachus pugnax)*, and migratory waterfowl and waders (NFFA, 1995).

CONCLUSION
Re-creation of intertidal wetland habitats is desirable on the basis that:

i) they represent a relatively rare, specialised and productive habitat;

ii) their wider importance (beyond designation for nature conservation reasons alone) is increasingly being recognised and valued; and

iii) reductions in quantity (extent) and quality (ecosystem health) of the present resource continue to be recorded

Monitoring suggests a combination of past and present engineering works, accelerated rising sea-level and reducing sediment loads have upset the equilibrium of coastal wetlands in the UK. The long-term outlook for intertidal wetlands, especially where sea walls prevent landward transgression, is not good.

Biophysical requirements of intertidal wetlands are increasingly understood and re-creation guidelines have been produced (Burd, 1995). However, to date, with the exception of a limited number of experimental projects, planners and managers have failed achieve the socio-economic prerequisites needed to take these projects forward.

The Danish system of land consolidation provides a potential solution. Projects, where this process has been used to initiate intertidal wetland re-creations, are now producing significant biodiversity gains. In each of these cases government finance has allowed the Danish National Forest and Nature Agency to acquire a "key property" which then acts as a pool for the voluntary redistribution of land among local farmers.

To achieve success decisions regarding land reallotment have to be achieved by voluntary consensus. The process is dependant on participation of local communities and the contribution of a professional mediator. Sidaway (1998) recognised the need for environmental consensus building and the pressing need to facilitate large scale managed realignment provides an ideal opportunity for these skills to be employed.

ACKNOWLEDGEMENT
The author would like to thank Bodil Ekner of the Danish Directorate for Development in Agriculture and Fisheries for his help in compiling this paper.

REFERENCES

Boorman, L.A., Goss-Custard, J.D. and McGrorty, S. (1989) *Climatic change, rising sea-level and the British Coast.* ITE Research Publication No.1, HMSO.

Burd, F. (1995) *Managed Retreat: A Practical Guide.* Campaign for a Living Coast. English Nature, Peterborough.

Carpenter, K.E. and Pye, K (1997) *Relative importance of changes in sea level and wind/wave climate on the stability of UK saltmarshes.* Fifth Symposium on the biogeochemistry of wetlands, Royal Holloway University of London 16-19 September, 1997. Unpublished.

Carter, R.W. (1989) Rising Sea-Level. *Geology Today.* **March/April** pp.63-67.

Collins, T., Leafe, R. and Lowe, J. (1997) *Sustainable flood defence and habitat conservation in estuaries: A strategic framework.* English Nature/Environment Agency.

Dixon, A.M., Leggett, D.J. and Weight, R.C. (1998) Habitat Creation Opportunities for Landward Coastal Re-alignment: Essex Case Studies. *Journal of the Chartered Institute of Water and Environmental Management* **12(2)** pp. 107-112.

DoE (1996) *Review of the potential effects of Climate Change in the UK.* UK Climate Change Impacts Review Group Second Report. HMSO, London.

EC (1998) *Report on the Progress of the Integrated Coastal Zone Management Demonstration Programme.* Communication from the Commission to the Council and European Parliament COM(97)744 final. Brussels 12 January, 1998.

Ekner, B. (1995) *Land Consolidation Schemes as a Tool for Nature Restoration.* In Proceedings of Nature Restoration in the European Union pp. 70-73. Ministry of Environment and Energy, The National Forest and Nature Agency Denmark.

MAFF (1993) *Coastal Defence and the Environment: A Guide to Good Practice.* Ministry of Agriculture Fisheries and Food and the Welsh Office.

NFFA (1995) *Restoration of sites in Special Protected Areas in Denmark.* ACE-project. Report No. 2242/91/09-1* Contract B191/91/SIN/8208. National Forest and Nature Agency, Ministry of Energy and Environment, Denmark.

NRA (1995) *A Guide to the Understanding and Management of Saltmarshes.* R&D Note 324. National Rivers Authority, Bristol.

NRA (1996) *Maintenance and Enhancement of Saltmarshes.* R&D Note 473. National Rivers Authority, Bristol.

Pethick, J. (1996) *Towards Integration: The Need for National and Regional Structures.* NCEAG Conference CZM: The New Agenda. London 4 November 1996. Unpublished.

Pye, K. and French, P.W. (1993) *Targets for coastal habitat re-creation.* English Nature Science 13. English Nature, Peterborough.

Shennan, I. (1993) Sea-level change and the threat of coastal inundation. *Geographical Journal* **159** pp. 148-156.

Sidaway, R. (1998) *Good Practice in Rural Development No. 5: Consensus Building.* Published by The Scottish Office Agriculture Environment and Fisheries Department for Scottish National Rural Partnership.

Sorensen, J. (1997) National and International Efforts at Integrated Coastal Zone Management: Definitions, Achievements and Lessons. *Coastal Management* **25** pp. 3-41.

Sorenson, E.M. (1994) *Agricultural Changes - Changing Agriculture: Some Danish Experiences and Perspectives of Multifunctional Rural Management.* Paper TS 708.2. International Federation of Surveyors XX Congress, Melbourne Australia March 5-12, 1994.

Tooley, M.J. (1989) The flood behind the embankment. *Geographical Magazine.* **November** pp.32-36.

Townend, I. (1997) *Realising the benefits of Shoreline Management.* RGS-IBG Seminar 29 October, 1997. "Enhancing Coastal Resilience: Planning for an uncertain future." Unpublished.

Coping with dynamic change on the coast – do we have the right regulatory system?

S.A. JOHN, Manager, Posford Duvivier Environment, Rightwell House, Bretton, Peterborough, UK, PE3 8DW (sjohn@posford.co.uk), and
R.N. LEAFE, Manager, External Relations Team, English Nature, Northminster House, Peterborough, UK, PE1 1UA (Richard.Leafe@English-Nature.org.uk)

INTRODUCTION

The system for regulating coastal planning and management in England is currently struggling to accommodate man-made change and does not really begin to address natural change. Much progress has been made in recent years to better co-ordinate the different legitimate uses of the coastal zone and to limit those activities which do not need a coastal location. Approaches to coastal defence planning have also improved significantly and the EU Habitats Directive has brought clear advantages for nature conservation. However, due to the variable quality of implementation and a lack of enforcement of non-statutory initiatives, real change on the ground has been slow in coming.

Over the last decade, English Nature and others have strongly advocated that, in order to provide a healthy living coastline, dynamic coastal processes must be able to take their course; providing benefits for both nature conservation and coastal defence. Yet, at times, the implementation of EC Habitats Directive in the UK appears to be in direct conflict with this aim, as it attempts to *preserve* conservation sites in their current form rather than *conserve* the interest features as part of a dynamic system. As a regulating tool the Habitats Directive has provided a much needed new balance to development decisions at the coast. However, experience of dealing with the Regulations has been mixed, with positive outcomes for development and nature conservation, but with complex bureaucracy along the way. Furthermore, it is apparent that meeting the requirements of environmental legislation alone will not achieve sustainable coastal management.

Ultimately, dynamic coastlines have to be allowed to function, and the regulatory system in England must facilitate this if coastal management decisions are to achieve sustainable results. This paper reviews recent cases that illustrate the dilemmas still faced in pursuing this aim, and suggests some reforms to increase progress toward a sustainable coastline. Cases illustrating current experience with the Habitats Directive and Coastal Defence Policy are reviewed under these headings.

THE HABITATS DIRECTIVE

The Conservation (Natural Habitat &c) Regulations, 1994, the implementing regulations for the EC Habitats Directive in the UK, have thus far delivered mixed benefits for coastal management. They have helped to mitigate potential harm to the environment by bringing the full weight of regulatory control to bear on developers proposing works within or adjacent to sites of European significance, that is candidate Special Areas of Conservation (cSAC) and Special Protection Areas (SPA). The recent port related channel deepening at Harwich is one example of the development control imposed by the Regulations. However, they have proved to be inflexible in dealing with on-going and inevitable change that occurs on the coast, for example, at Brancaster in North Norfolk, the implementation of a managed realignment solution has been delayed while issues around how to conserve all of the conservation features of interest at the site are resolved.

The Port of Felixstowe V the Stour and Orwell Estuaries SPA

In 1997, the Harwich Haven Authority proposed to deepen the approach channel to the Haven Ports by 2m, to -14.5m CD. These works, now underway, are taking place immediately adjacent to the Stour and Orwell Estuaries SPA. Given this, and because it was predicted that the deepening would have an adverse affect on the integrity of the SPA (HR Wallingford and PDE, 1998), a number of mitigation and compensation measures are being undertaken in conjunction with the works. The outcome is positive in terms of nature conservation gain, in that a sustainable solution has been achieved as a result of the requirements of the Habitats Directive in combination with the financial ability of the ports industry to deliver the solution. In fact, the long-term effect of the sediment replacement proposed will be to mitigate for previous dredging operations and, to some extent, background change.

Without either part of this equation – regulatory enforcement and industry finance – this solution would not have been achieved. However, it took two years and a substantial financial outlay. Because of the requirement for all components of the scheme to be considered as part of one project, sixteen scheme components had to be assessed and seven environmental reports produced (as detailed below), for four different regulatory bodies:

- Approach Channel Deepening Environmental Statement (ES) – addressing the potential impact of the deepening, the channel, disposal and use operations, and ongoing maintenance dredging.

- Approach Channel Deepening Appropriate Assessment – specifically addressing the potential affect of the works on the integrity of the SPA and expanding on the mitigation measures proposed in the ES.

- Offshore Disposal Ground Supplementary Environmental Statement – because the ES recommended that an alternative disposal ground was sought for the future disposal of maintenance material, the DETR required that the new ground was found, assessed and approved (by MAFF) prior to the consent being awarded for the deepening.

- Approach Channel Deepening Mitigation and Monitoring Strategy – in light of the prediction that the deepening would cause an immediate loss of 4ha of the inter-tidal habitat (due to a change in the tidal range) and (on average) the loss of a further 2.5ha per annum, a strategy was developed to mitigate this impact and to monitor its outcome.

- Sediment Replacement Appropriate Assessment – the potential effect on the SPA of the measures recommended as mitigation for the predicted annual erosion of 2.5ha of inter-tidal also required assessment.

- Trimley Marsh Supplementary Environmental Statement etc. – managed retreat was proposed at Trimley to compensate for the immediate 4ha loss and a further, precautionary, 12.5ha (representing 5 years of unmitigated loss). Again, approval for the retreat had to be in place prior to consent being awarded for the deepening and this required planning permission, environmental assessment, appropriate assessment and Food and Environment Protection Act (FEPA) consent. (Other compensation sites were also investigated and assessed).
- Beneficial Use Appropriate Assessment – to meet the requirements of FEPA, beneficial uses had to be sought for the shingle, sand and silts arising from the works and, in turn, the potential affect of placing such material within the site on the integrity of the SPA required assessment.

Given the complexity of the scheme and the number of different parties involved, reluctance on the part of any one 'competent authority' to grant consents for its part of the process, without being aware of the position of others, was also met.

The consent process for the Harwich Haven Authority Approach Channel Deepening illustrates that the current regulatory system is threatened by collapse under its own administrative weight. Furthermore, the Haven Authority were tied into a very expensive lease agreement with Trinity College for the retreat site at Trimley Marsh because of the requirement to obtain consent for, and have in place, all of the proposed mitigation measures prior to receiving Coast Protection Act consent for the deepening. Pressure on time in this case meant that more cost effective options could not be investigated and led to the purchase of a site which the Environment Agency have since revealed would have been subject to a managed retreat coastal defence policy in due course. *"This will surely make the new intertidal habitat, at over £100,000 per annum rental, the most expensive piece of mud in the world!"* (*G. Steele*, May 1999).

It is evident that we need to plan projects in order to ensure that this money is spent on environmental gain rather than land agents fees. This requires a strategic approach in order to make expenditure commensurate with environmental enhancement. Such a strategic approach should also allow for mitigation adjacent to European site boundaries. The dynamic nature of the coast means that habitat or sediment replacement, for example, adjacent to SPAs/cSACs can contribute to the functioning of the system and, therefore, the integrity of the site. In addition, due to the uncertainty associated with coastal processes, our approach would benefit from the acceptance of some risk associated with mitigation, however, this would need to be validated by monitoring and revisited if required performance levels were not achieved, and conditioned in law to ensure that the requirements of the Habitats Directive are ultimately met.

In the case of Harwich, because the retreat site at Trimley Marsh was outside the Stour and Orwell Estuaries SPA and was to be retreated after the works had begun, the Haven Authority eventually had to prove Over-riding Public Interest (OPI). This established a worrying precedent for smaller operators who, if they can not mitigate their proposals within the site boundaries, will not be able to prove imperative reasons of OPI. Clearly, in many cases of small scale change, the works proposed should not proceed within a designated site and their cumulative effects, with others proposals, must be considered. However, change will not always lead to damage, and the current regulatory may lead to development being constrained that has the potential to be sustainable.

A simpler, more streamlined decision-making process is required. This would help to relieve the pressure on the regulators and would provide the scheme promoters with a quicker decision that, although it may not be favourable, would limit their level of investment. Such a solution could entail decision-making through regional coastal regulatory groups. The regional level is geographically appropriate for coastal planning, given the scale of operation of physical processes (this would also take advantage of the new governance structure emerging for English regions). Such a group could assemble the relevant competent authorities for any one coastal area (e.g. DETR, MAFF, the local authority, the navigation authority and English Nature), based on a standing committee, to arbitrate over development decisions based on structured representations from scheme promoters.

Brancaster Flood Defence Scheme V the North Norfolk Coast cSAC and SPA
The main impact on European sites at the coast is likely to result from the implementation of coastal defence policies (contained in Shoreline Management Plans) is the loss of habitat due to 'coastal squeeze'. That is, where fixed coastal assets prevent the landward rollback of inter-tidal and littoral habitats or, alternatively, where retreat threatens saline lagoons and freshwater wetlands. Ensuring that the ecological requirements of European sites are met in the medium to long term, requires that losses are made good by gains through the restoration or re-creation of new areas. Managing for loss within the confines of designated site boundaries, however, could further compromise other interests of European importance. Therefore, to allow for dynamic coastal change, future habitat re-creation is likely to be necessary next to existing SPAs and SACs.

At Brancaster, a seawall protects an area of designated grazing marsh from the sea. This seawall is being undermined, threatening the site with inundation and thereby the loss of the SPA freshwater conservation interest. Holding the line, however, will damage the coastal cSAC due to coastal squeeze. A scheme is therefore being developed that will protect the grazing marsh *in situ* while achieving partial retreat. In other cases, however, such a solution frequently will not be achievable, introducing an urgent requirement for habitat restoration and/or re-creation elsewhere.

The situation at Brancaster (and nearby Cley) led to loss and gain habitat accounts being developed for all coastal cells in England and Wales, based on 'best-guess' coastal change and defence scenarios for the next 50 years (Lee, 1998). The premise being that mechanisms should be developed to achieve 'no nett loss' of conservation interest. In line with the Habitats Directive, a presumption also exists in favour of protecting European sites *in situ*. However, it is accepted that it will not always be possible to conserve that interest in the same location, especially when doing so would result in damage to another European site (Collins and Beardall, 1999).

Where such conflict exists, the proposed decision-making mechanism for coastal management is Coastal Habitat Management Plans (CHaMPs). These plans are intended to provide a framework for managing European sites that are located on or adjacent to dynamic coastlines. English Nature, the Environment Agency and MAFF have jointly proposed that CHaMPs will be used in circumstances where the conservation of all of the existing interests within a site complex *in situ* is not possible, due to proposed changes in coastal defence options or coastal process induced changes to the shoreline. Their primary function is to act as an accounting system to record losses and gains to the habitats and species of European importance subject to shoreline change, and to set the direction for habitat conservation measures to address net losses in any one functional area.

In this context, the Habitats Directive appears to favour the maintenance of the *status quo*, potentially tying the Environment Agency (in the case of Brancaster) to expensive and unsustainable coastal defence and adversely affecting coastal habitats that are dependant on the continuation of coastal processes. The avoidance of 'nett loss' from an arbitrary point in time will create a coastal ecosystem fixed in an 'as designated' state (circa 1995). Given our potential to manage (and manipulate) the coastal environment, it is apparent that we should focus on defining and achieving the best habitat balance we can, rather than a prescribed one. Again, this promotes that need for a more strategic approach to coastal management.

COASTAL DEFENCE POLICY
The implementation of Coastal Defence Policy, under the 1949 Coast Protection Act, can have two outcomes for the potential beneficiaries of a proposed defence scheme. They are either big winners, if the scheme is built, or they risk losing everything. Often caught in the middle of this very public debate, is nature conservation and other interests such as landscape, recreation and archaeology.

Castle Haven proposed Coast Protection Scheme V BAP priority habitats
The Isle of Wight Council is proposing to construct, under the Coast Protection Act, a defence for approximately 20 properties and associated infrastructure at Castle Haven on the south coast of the island. The defence, comprising some 600m of rock revetment, slope re-grading and drainage works, if constructed, would damage a rare ecological habitat, known as Maritime Cliff and Slope, which constitutes a priority habitat under the UK Biodiversity Action Plan (BAP).

The principle behind the Biodiversity Action Plan for these types of habitats is one of no nett loss. In the case of the Maritime Cliff and Slope Habitat Action Plan, no nett loss is essential given the state of decline of the resource, largely as a result of cliff stabilisation works for coastal defence schemes. MAFF, the Environment Agency and English Nature are currently discussing how to operationalise such a policy, including opening up cliff sections where old defences are now redundant.

The dilemma at Castle Haven is clear, either the residents of Castle Haven receive their 'belt and braces' protection scheme and the Maritime Cliff and Slope habitat becomes a little rarer, and Government a little further away from meeting its BAP commitments, or the properties are lost and the habitat remains. Partial defence, or options to slow the rate of erosion, are unlikely to satisfy either property owners concerns or safeguard the rarest components of the habitat that depend on instability. An alternative option is the planned, phased removal of properties as the threat becomes reality; supported by appropriate financial mechanisms. The inflexibility of the current regulatory system, that adopts an all or nothing approach, does not allow such a middle course to be taken.

East Head Coastal Defence Strategy V Navigation interests
East Head, a coastal spit and SSSI, forms the eastern side of the mouth of Chichester Harbour. The thin neck, or hinge, of the spit is currently the subject of a privately funded proposal to provide protection through re-enforcement, via a rock revetment placed on its land-ward face. The main driver behind the proposal concerns the perceived threat to navigation resulting from a permanent breach at the neck, leading to a new entrance to the Harbour developing. Indeed, geomorphologically, the current mouth of the harbour is too small and is attempting to increase through erosion.

If the proposal goes ahead it would represent another example of piecemeal coastal defence. It would not contribute to a sustainable coastal defence strategy for East Head and the shoreline adjacent and, through time, could potentially increase the likelihood of a permanent breach by interfering with the natural process. It is widely acknowledged that we do not know enough about the geomorphological evolution and operation of this system, yet responsibility for researching the issue at this time is unclear.

What is lacking in this case is an arbitrator for the current decision, able to take a view of the overall merits of the case and place it within a wider strategy for the management of this part of the coast as a whole.

CONCLUSIONS
As the above examples illustrate, given the current regulatory framework, the realisation of a sustainable coast in England is still some way off, despite recent non-statutory planning initiatives and the tightening of the conservation requirements through the Habitats Regulations.

Nonetheless, we are significantly ahead of many of our European neighbours in striving for sustainability. Our policy developments over the past few years, particularly in the field of coastal defence, are substantial and significant. However, we must not rest on our laurels, but release the full potential of all that has gone before, recognising that making the final few moves may be the hardest yet. Some key issues, in need of urgent debate, that flow from the case studies presented here, are summarised below:

Key issues for Habitats Directive implementation in coastal areas
Our interpretation of the Habitats Directive needs to cater better for dynamic change. It is important to avoid being taken down an increasingly preservationist path in order to meet the letter of the UK implementing Regulations. Such a course will not ultimately help achieve the Directive's aim of 'favourable conservation status' for coastal habitats and species. The outcome at Harwich, that mitigation can only occur within the current site boundaries, jeopardises the potential to reach sustainable solutions elsewhere, which may offer better safeguards in the long-term for conservation interests. Was this really the intent when the Directive's implementing regulations were drafted? Is it time for an amendment?

It is hoped that the Environment Agency/English Nature joint LIFE sponsored project 'Living with the Sea - Managing Dynamic Coastlines' will draw attention to this issue and make recommendations for solutions based on practical examples (through CHaMPs).

Equally, recent cases have demonstrated that the bureaucracy surrounding the consents procedure is complex and difficult to navigate a way through for large developments. There are now pressing arguments for a rationalisation of regulatory authority on the coast, providing a 'one stop shop' for consents applicants. There remains only one untried solution, primary legislation for Coastal Zone Management.

Key issues for Coastal Defence Policy

The Coast Protection Act, 1949, has served its time. It needs to be replaced with new legislation that takes a broader view of the requirements of coastal management. Mechanisms to avoid the big winners and losers of the present system should be considered. Crucially, financial mechanisms to enable relocation of threatened assets, rather than protect or bust, are urgently needed. This is not another simplistic call for compensation, but a plea for a much needed serious examination of options that could facilitate the removal of public and private assets of limited value, which currently prejudice the delivery of sustainable defences and living coastlines. The groundwork for identifying the potential size, and broad cost/benefit, of such a provision is now nearing completion through Shoreline Management Planning.

We should prepare to make this next major step in the evolution of coast management now, pre-empting any legislative requirement that may ultimately come from the European Community.

REFERENCES

Collins, T. and Beardall, C. (1999), Coastal Defence and the Habitats Regulations: plans and actions. *34th MAFF Conference of River and Coastal Engineers*, 30th June to 2nd July 1999, Keele University, 6.4.1-6.4.6.

EC Demonstration Programme (1999), *Towards a European Integrated Coastal Zone Management (ICZM) Strategy: General Principles and Policy Options - a reflection paper.* EU Demonstration Programme on Integrated Management in Coastal Zones 1997-1999.

House of Commons: Sessions 1991-92 (1992), Environment Committee – Second Report, *Coastal Zone Protection and Planning - April 1992*, HMSO.

HR/PDE (1998), *Harwich Haven Approach Channel Deepening – Environmental Statement*, Report EX 3791, January 1998.

Leafe, R.N., Pethick, J.S. and Townend, I. (1998), Realising the Benefits of Shoreline Management. *The Geographical Journal, Vol. 164 No. 3, November 1998, pp. 282-290.*

Lee, M. (1998), *The implications of future shoreline management on protected habitats in England and Wales.* Environment Agency R&D Technical Report 150.

Murby, P. and Pullen S. (1999), A Working Criteria for Environmental Acceptability. *34th MAFF Conference of River and Coastal Engineers*, 30th June to 2nd July 1999, Keele University, 6.1.1-6.4.7.

PDE/HR (1998), *Harwich Haven Approach Channel Deepening – Appropriate Assessment*, May 1998.

PDE/HR (1998), *Harwich Haven Approach Channel Deepening – Mitigation and Monitoring Package*, October 1998.

Steele, G. (1999), *The Port of Felixstowe, Development and the Habitats Directive: Conflict or Consensus?* Posford Duvivier Environment Seminar, May 1999.

A sea defence strategy for Salthouse on the north Norfolk coast

S J HAYMAN
Environment Agency, Anglian Region.

INTRODUCTION
The North Norfolk coastline is recognised as an area of exceptionally high nature conservation value. Much of the coastal frontage is low lying with large areas of freshwater marsh which were reclaimed from the sea between the 16th and 18th centuries. The Environment Agency has inherited responsibility for the sea defences which protect these areas against tidal inundation. The Agency's objective is to manage these defences so as to minimise the risk of flooding to people and property, having due regard for the needs of conservation, local communities and the wider coastal system.

As the coastal marshes are of such high environmental and ecological significance there is an understandable desire to see them safeguarded at all cost, and to resist any form of change. Indeed, a large part of the area is designated under current EC legislation which imposes a responsibility to ensure that there is no net loss of habitat. It needs to be appreciated, however, that in a dynamic coastal situation it is inevitable that there will be some environmental change. The challenge is to move towards strategies for the management of this coastline which emulate the natural processes with, as far as practicable, no overall detrimental effects on the diverse coastal habitats.

This paper describes work which is currently being undertaken by the Agency and their consultants Halcrow to develop a sustainable sea defence strategy for the length of coast between Cley and Kelling which will meet the criteria of social and environmental acceptability, economic feasibility and engineering soundness. The area at risk of flooding is fronted by a 5.5 kilometre quasi-natural shingle ridge with a clay embankment along the estuary of the River Glaven providing the defence on the western flank (see Figure 1).

TOPOGRAPHY
The shoreline of the North Norfolk coast between Sheringham and Blakeney Point is characterised by a natural upper shingle ridge beach. Between Sheringham and Kelling Hard the beach is backed by glacial cliffs, but to the west of Kelling Hard the single ridge fronts an area of freshwater formed by enclosure of the saltmarshes.

These coastal marshlands are separated from the rising ground of the Cromer Ridge to the south by the A149 coast road and the settlements of Cley and Salthouse. The river Glaven runs by Cley on its route to the sea, but is then deflected westwards for 6 kilometres by the presence of the

PAPER 7 : HAYMAN 67

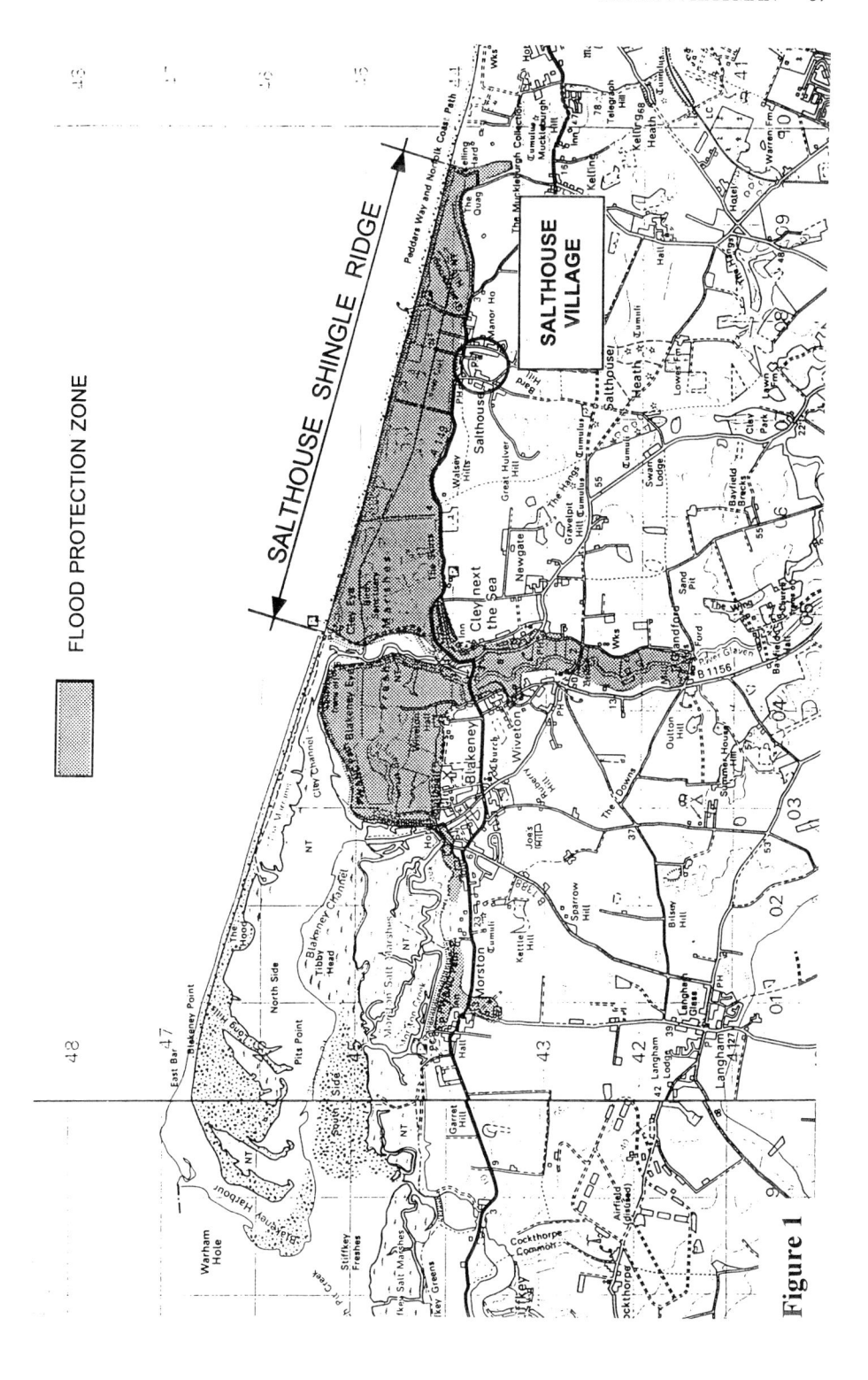

Figure 1

sand and shingle spit leading to Blakeney Point. The whole of the coastal plain and lower reaches of the Glaven valley are below the 5.0 metre contour, and therefore within the tidal flood risk area.

INFRASTRUCTURE

Between the 13th and 16th centuries the 'Glaven Ports' of Cley, Blakeney, Morston, Wiveton and Salthouse were busy and prosperous ports. Ships of up to 150 tons would export Norfolk's farm produce and return with cargoes of wines, spices, linen and other luxury goods. Reclamation of the saltmarshes for grazing accelerated siltation of the channels and harbours and this, together with the advent of the railway, brought about their decline as centres of maritime activity until by 1800 seaborne trade and almost ceased.

From the early 20th century onwards the villages have increasingly become holiday and retirement centres with many visitors attracted by the nature conservation interest and rich landscape of the coastal marshland. Both Salthouse (population 192) and Cley (population 475) lie on the A149 coast road which is the main route from Cromer to Hunstanton.

COASTAL GEOMORPHOLOGY

The coastal ridge to the west of Kelling Hard can be divided into two process units, the length to Cley Coastguards which is mostly shingle, and the section which is predominantly sand between Cley Coastguards and Blakeney Point. The geomorphology of the ridge and its adjacent coastal systems is complex and requires careful assessment before future management decisions can be taken. The Agency therefore commissioned the Coastal Research Unit at Cambridge University (CCRU) to assess the natural processes and forces operating on this shoreline, research the historical development of the ridge and examine the environmental issues pertaining to a range of future management options.

CCRU reported on their investigations in March 1998 and concluded that the shingle section of the ridge is based upon an underlying morainic ridge formed during the last glacial period along the eastern limit of the advancing ice. It would appear that the shingle ridge was first formed after the glaciation when the rising sea level swept before it material derived from erosion of the dry seabed. This being the case, it is likely that no significant contribution to the total volume of the ridge is being made by present day processes.

The ridge has been moving landwards at approximately 1 metre per year since at least 1630. The presence of a series of glacial till mounds, known as the Eyes, may have acted to retard the landward movement of the ridge. These eyes will be eroded away over the next 150 to 500 years and this could result in an accelerated rate of retreat of the ridge.

ECOLOGY

To landward of the shingle ridge there are three significant areas of freshwater habitat, the Cley Marshes, Salthouse Marshes, and the Blakeney Freshes, all of which are protected from regular saline inundation. All three areas are of high environmental and ecological value, comprising reclaimed coastal marshes with fresh/brackish ditch networks, extensive areas of reedbed and some shallow water lagoons.

The major nature conservation interest of the marshes is the internationally and nationally important numbers of breeding and wintering birds. The ornithological importance lies in the vast numbers of birds which use the area as well as the presence of rare species such as bittern, bearded tit and marsh harrier within the reedbeds. In several area the freshwater levels are managed for breeding waders, mainly redshank and lapwing and wildfowl. Significant numbers of brent geese, teal and widgeon use the sites and avocets make use of the scrapes at Cley and the bare and/mud areas at Salthouse. The area is a major tourist attraction, with birdwatching a particular feature (the Norfolk Wildlife Trust Cley Reserve attracting over 100,000 visitors each year).

The importance of the area in nature conservation and landscape terms is recognised by the following national and international designations:

- North Norfolk Coast Special Protection Area (SPA) designated under the EC Directive on the Conservation of Wild Birds;

- The North Norfolk and Gibraltar Point Dunes cSAC, designated under the EC Council Directive on the Conservation of Natural Habitats and of Wild Fauna and Flora: The Habitats Directive;

- The Wash and North Norfolk Coast Marine candidate Special Area of Conservation (cSAC);

- The North Norfolk Coast Site of Special Scientific Interest (SSSI);

- The North Norfolk Coast Ramsar Site, Convention on Wetlands of International Importance Especially as Waterfowl Habitat;

- The North Norfolk Coast Area of Outstanding Natural Beauty (AONB);

- The North Norfolk Coast Biosphere Reserve; and

- The North Norfolk Heritage Coast.

FLOOD PROTECTION
The sea defences around the village of Cley were upgraded between 1993 and 1995 by the construction of a new outer ring embankment across the flood plain of the River Glaven just to the north of Cley Mill. Where the embankment crosses the main river channel, a pair of tidal gates has been provided which are closed automatically for the duration of the surge tide. This scheme, which protects 122 properties, has been designed to withstand a repeat of the 1953 event (a return frequency of approximately 1 in 100 years).

To the east of Cley, the low-lying area at risk of flooding includes 35 properties in Salthouse and Cley villages, the A149 Coast Road and 300 hectares of reclaimed grazing marshes. This area is totally dependent upon the protection provided by the managed length of shingle ridge between Cley Coastguards and Kelling Quag.

The Agency has continued a reprofiling maintenance programme initiated by its predecessors on the shingle ridge to enhance its sea defence properties. This normally involves bulldozing shingle from the lower beach to maintain a crest elevation considerably in excess of the natural level. Emergency works are also undertaken when necessary to restore the integrity of the ridge following damage by severe wave action during storms. The scope of these works is restricted by the limited sediment input along the frontage and it should be noted that this intervention has arrested the rate of landward advance of the ridge, and over steepening the seaward face can result in increased wave energy at the crest.

The last flood event to seriously affect properties in Cley and Salthouse villages was on 31st January 1953 when one person was drowned and several houses were flooded with some being damaged beyond repair. More recently, the shingle ridge has been overwhelmed during storm surges in 1976, 1978, 1993 and 1996, with resultant flood damage to both the marshes and coast road on each occasion.

A Shoreline Management Plan for North Norfolk was published in July 1996. With reference to the maintenance of the shingle ridge it states:

> "....with the high environmental value of the marshes the annual expenditure on the reprofiling is justifiable. However, if the supply of shingle to the foreshore continues to reduce then an alternative means of locally protecting Salthouse should be provided."

MANAGEMENT POLICY OPTIONS

The Shoreline Management Plan defined the primary coastal defence objective for this section of coast as: -

"Continue to provide flood defence to Cley and Salthouse either through maintenance of existing flood defences, including maintenance of the shingle ridge, or through the construction of new defences if the shingle ridge is allowed to retreat.
Protect Salthouse and Cley Marshes until these marshes can be recreated elsewhere."

It recommended an evaluation of the full range of policy options along this frontage, which can be summarised as follows: -

(i) **Do Nothing**
If no further work was to be carried out on the sea defences then the shingle ridge would be breached and properties in Salthouse and the eastern end of Cley would be at increased risk of flooding. There would be total loss of the freshwater habitat and the A149 coast road. This option is included because it provides the yardstick for assessing the other options.

(ii) **Continue Re-profiling Shingle Ridge and Repair Breaches**
It is considered that this policy is not sustainable in the medium to long term because there is insufficient shingle available to maintain the ridge to an adequate profile and crest elevation. The ridge will become increasingly permeable and unstable during storm events, thereby reducing its efficiency as a flood defence structure.

(iii) **Large Scale Shingle Nourishment**
There are very few economically or environmentally viable sources of suitable material and this would be a very costly option. There could be significant environmental impact on the donor site, either terrestrial or marine.

(iv) **Full Scale Retreat**
This would lead to the loss of freshwater marshes and the coast road would be subject to periodic flooding. Any embankment substantial enough to provide effective protection against major surge tides would have a significant impact on the visual amenity of the area.

(v) **Partial Set Back of the Sea Defence Line**
This option allows for the construction of a new secondary embankment landward of the shingle ridge extending the full length of the frontage. The embankment would be designed to withstand overtopping in extreme conditions, although the residual shingle ridge would act as a primary wavebreak. Future maintenance of the ridge would then by much reduced allowing it to resume a more natural profile.

Before developing any of the options in detail the Agency decided to embark upon a participatory and interactive consultation process with local councils, conservation bodies, landowners and residents. The intention behind involving the full range of interested parties at such an early stage was to generate a sense of involvement and shared ownership in the eventual outcome. Through this process the consultees came to appreciate the wide range of constraints (e.g. legislative, financial, technical, etc) which faced us in reaching an affordable and sustainable solution.

The consultation was led by the Centre for Social and Economic Research in the Global Environment (CSERGE) at the University of East Anglia. Meetings of a "coastal forum" were arranged to afford consultees the opportunity to express their concerns and aspirations and to identify potential areas of conflict. The serious damage caused by the flood event in February 1996 was still fresh in the memory and there was widespread acceptance that the present reliance on management of the shingle ridge was no longer effective and there was a need for change. It emerged from these initial consultations that there was a strong local consensus in favour of the partial set back clay embankment, and the Agency undertook to commission a study to investigate the best way of implementing this option. All participants in the forum expressed the wish to continue their involvement throughout the future scheme development.

ENVIRONMENTAL ASSESSMENT
It is clear that maintaining the status quo as far as the wildlife interest of the Cley and Salthouse Marshes is concerned is not a viable option. To 'do nothing' would obviously result in the eventual deterioration of the whole freshwater environment due to saline inundation whereas it appeared that the provision of a secondary defence to maintain the majority of these important habitats should be environmentally beneficial overall as well as providing improved flood protection for the local community. The Agency therefore engaged Halcrow as its consultant to appraise the options through a statutory Environmental Assessment leading to the production of an Environmental Statement.

The work carried out by Halcrow in preparing the Environmental Statement included:

- Collation of baseline data for existing land use and management, archaeological sites, etc

- Undertaking a Phase 1 Habitat Survey, Landscape and Visual Assessment and a coastal processes study

- Extensive consultation involving the distribution of a scoping document, site visits with consultees and public meetings

- Impact assessments for a range of options with a comprehensive assessment of the preferred option

- Identification of mitigation and enhancement, and the

- Production of an Environmental Action Plan

An initial consultation document was produced to inform consultees of the Agency's objectives, the background to the assessment and the preliminary proposals. Consultees were invited to provide an initial response and/or to be kept informed as the project progressed. In addition to the wide circulation of this document at the scoping stage, the assessment was advertised in the local press in order to provide all members of the public an opportunity to comment or request further information.

The Environmental Statement initially confirms Partial Setback as the preferred option and then proceeds to refine the scheme with respect to embankment alignment and profile, and material sourcing. The proposed embankment is approximately 3.5m in height, with a 3 metres wide crest and a 24m base width, extending from the beach road at Cley Coastguard to high ground east of Salthouse (see Figure 2). It is inevitable there will be changes to the ecology of the area, and these are summarised in the Environmental Statement as follows: -

> "There will be a loss of grazing marsh, reedbed, openwater and saltmarsh through the footprint of the embankment, however, these will be overall improvements to these habitats. 78% of the grazing marsh, 40% of open water and 88% of reedbed will become increasingly fresh due to a reduction in saline inundation. Approximately 36.1 hectares of saltmarsh south of the embankment will revert back to freshwater grazing marsh. There will be the creation of a further approximately 14.2 hectares of reedbed and the lowering and subsequent improvement of approximately 12.1 hectares of reedbed. Approximately 11.9 hectares of additional lagoons will also be created north of the embankment. Investigations have concluded that the water budget for the site will be adequate for the newly created habitats. Improvements in water level management and drainage will allow the optimum environmental conditions for the study area to be realised."

Proposals for a monitoring programme, during the construction phase and on a long-term basis, are made in the Environmental Statement which also recommends separate studies to develop a Water Level Management Plan and a Beach Management Plan. Both of these have subsequently

PAPER 7 : HAYMAN   73

Aerial photograph of the Salthouse frontage showing the proposed clay embankment and sources of clay

Figure 2

been completed.

ENVIRONMENTAL LEGISLATIVE FRAMEWORK

The Agency's principal aim, defined in the Environment Act 1995, is that "in discharging its functions to protect or enhance the environment it will contribute towards attaining the objective of achieving sustainable development." Flood defence and sustainable development are closely linked, and the Agency favours solutions which take account of natural processes and safeguard wildlife habitats whilst having regard to the economic and social well-being of local communities.

It is on this basis that the Agency is promoting the partial retreat option at Salthouse and an application for planning consent has been submitted to the local authority. This application is supported by the Environmental Statement in accordance with the requirements of statutory instrument 1199, the Town and Country Planning (Assessment of Environmental Effects) Regulations 1988.

The legislative situation is more complicated, however, through the legal protection that conservation sites enjoy under The Conservation (Natural Habitats, &c.) Regulations 1994 which transpose the EC Habitats Directive into law in Great Britain. The Regulations impose an obligation to avoid damage to habitats and species of European importance or to provide compensatory habitat if there is a negative impact, even where the change in the ecological balance is a result of natural processes. We have the perverse situation, then, where legislation intended to further the conservation of natural habitats may make it more difficult to implement sustainable shoreline strategies.

On a positive note, the Agency has been working closely with English Nature to address this dilemma through the development of a North Norfolk Coastal Habitat Management Plan (CHaMP). The aim of CHaMPs is to set down a framework for balancing losses and gains of features in SACs, SPAs and Ramsar sites on dynamic coastlines over a 50-year period, thereby fulfilling our statutory obligations under the Habitats Regulations. Collins and Beardall, 1999 provides a fuller description of the Habitats Regulation and CHaMPs.

It also needs recording at this point that the construction of the secondary embankment and associated drainage works at Salthouse is estimated to cost £3.5 million. By the established MAFF project appraisal guidance criteria used when determining applications for grant aid on flood defence schemes the economic justification for this scheme is, at best, marginal. This should not now present an obstacle to the progress of the works, however, as the Government announced in July last year that the usual arrangements would be adjusted, where necessary, to accommodate the funding of measures necessary to protect the ecological integrity of sensitive sites of European importance.

THE CONSULTATION PROCESS

Reference has been made previously to the participatory and interactive approach adopted to consultation on this project, but this warrants a further brief comment with respect to the implementation of strategic coastal management policies in general.

As shoreline managers we might have been feeling fairly satisfied that all our SMPs were in place when along comes the House of Commons Agriculture Select Committee on Flood and Coastal

Defence to challenge whether all these worthy strategic plans and priorities are being translated into "positive action on the ground." On the evidence of consultations in respect of Salthouse proposals, the progress made towards a strategic approach to the management of the shoreline is not widely appreciated and we certainly have some way to go in convincing residents and interest groups who are focussed on local issues that there is a need to consider the bigger picture. There are many who believe our goal should be preservation of the coast in its present form, and advocate more intervention in the natural processes - not less. Coupled with this is the suspicion that references to 'soft' engineering and working with nature are code for saying that we are going to cut back on our maintenance budgets.

The involvement of the community as well as conservation organisations and local authorities in the decision making process at Salthouse has worked well. Forums arranged in the village at which consultees with differing aspirations were encouraged to put forward their viewpoints openly at the start did much to resolve conflicts which may otherwise have surfaced late in the development of the scheme.

Establishing a trusting partnership with the stakeholders takes much time and effort. The process has been ongoing at Salthouse for two years on a project covering 5 kilometres of coastline out of a total 4500 kilometres for England and Wales, so there are obvious implications in terms of cost and resources. The potential benefits for the long term more than justify the investment.

CONCLUSIONS AND SUMMARY

The Agency, following extensive investigations and consultations, is promoting a partial setback scheme at Salthouse as the most effective and affordable long-term sea defence management option. We are hopeful that the necessary approvals to proceed with the works will be in place before the next summer construction window, but this is by no means certain. Whatever the outcome, however, there are some conclusions to be drawn from the work to date.

Sustainability of sea defences is a key area of concern for the Agency, and also for the nature conservation organisations. Working as partners with common objectives we should be able to find a way through the maze of legislation and succeed in implementing integrated coastal strategies. There is an obvious requirement to comply with our statutory obligations, but there is also a need for some political pragmatism in how we apply the various Regulations to ensure that these do not get used as an excuse for inaction whilst our plans founder and the potential benefits for people, property and the environment are lost.

It is important to appreciate the importance of involving the wider public who live and work on the coast through the whole life cycle of plan implementation. The support which is mobilised when consultees feel a sense of shared ownership for the proposals can provide an essential catalyst in achieving a successful outcome. We have experienced the benefits of introducing an interactive consultative style in the decision-making process at Salthouse, but the resource implications should not be underestimated.

ACKNOWLEDGEMENTS

The author gratefully acknowledges the input to the study provided by Lucy Bazeley and Gareth Heatley at Halcrow, Rosie Ward and Tim O'Riordan (CSERGE) and John Pethick. Thanks also go to Robert Runcie and Paul Miller for comments during the preparation of the paper.

REFERENCES

Collins T and Beardall C 1999 — Coastal Defence and the Habitats Regulations: Plans and Actions. Paper to 34th MAFF Conference

EC Council Directive 92/43/EEC: — The Habitats Directive

Halcrow 1998 — Salthouse Flood Protection Environmental Statement

House of Commons 1998 — Flood and Coastal Defences Agriculture Select Committee Report

MAFF 1993 — Project Appraisal Guidance Notes for Flood and Coastal Defence

Mouchel L G 1996 — North Norfolk Shoreline Management Plan

O'Riordan T and Ward R — Building Trust in Shoreline Management CSERGE Working Paper GEC 97-11

Pethick J 1997 — Coastal Research Unit, Cambridge University Cley to Kelling Environmental Investigation

Analysis of the Long-term Variations in Offshore Sandbanks

Bin Li[1], D.E.Reeve[2], N. Thurston[1]
[1]Halcrow Group Ltd, Burderop Park, Swindon, Wiltshire, SN4 0QD, UK
[2]School of Civil Engineering, University of Nottingham, Nottingham, NG7 2RD, UK

ABSTRACT

This paper describes results from a study funded by MAFF to develop statistical methods for analysing changes in coastal morphology over the long term. Statistical eigenfunction and GIS methods for analysing long-term changes in coastal morphology are applied to the region of sandbanks several kilometres offshore of Gt Yarmouth. Results suggest that there are significant recurrent morphological variations, which occur over periods longer than the current planning horizons adopted in, for example, SMPs.

INTRODUCTION

Events, such as the major cliff falls at Beachy Head in early 1999 have made headlines in the national press. As a consequence, the general public in the UK has become increasingly aware of the threat posed by coastal erosion and flooding. This increased awareness has brought new demands for improved or new sea defences to protect homes, properties and infrastructure. Although the implementation of Shoreline Management Plans (SMPs) has begun the process towards implementing a nationally consistent planning framework for coastal zone development, our understanding of coastal processes, in particular over the long term remains wanting. This research deficiency is a problem as coastal processes is the crucial ingredient on which successful planning process depends. To therefore supplement the many methods of assessing the short-term (weeks to a year) coastal response to physical processes, several recent research studies have focussed on methods for assessing the medium to long-term (decades to a century) changes in coastal morphology (Sims, 1995; Raper et al 1997).

In this paper, results from research into the long-term morphological changes of the Gt Yarmouth sandbanks, East Anglia are described. They have considerable importance as they protect the region's beaches from wave action; their movement can prove hazardous to navigation and they are an important breeding ground for grey seals. Earlier studies have postulated recurrent behaviour in the sandbank configuration (Robinson, 1960; Halcrow, 1991; Reeve 1995).

Coastal Management: Integrating science, engineering and management, Thomas Telford, London, 2000

DATA

Historic bathymetric charts of the area covering the period from 1846 to 1992 have provided an extremely good record both in terms of their spatial resolution and temporal coverage. The area considered is approximately 12km wide and 35km long. Historical data were abstracted from sixteen bathymetric charts for dates between 1846 and 1992, for an area that extends from Winterton in the north to Benacre in the south. Figure 1 shows the study area and the bathymetry for 1992.

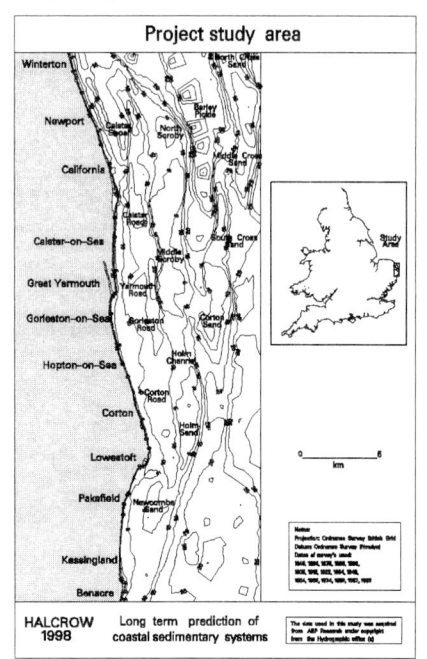

Key features of the sand bank system are:

- Holm Sand, 3km due east of Lowestoft;
- Corton Sand, 5km due east of Gorleston-on-Sea;
- Middle and North Scroby Sands lying 2km offshore and running almost parallel to the mainland between Gt Yarmouth and Newport;
- South, Middle and North Cross Sands running almost parallel to the Scroby Sands but approximately another 3km further offshore;
- Caister Shoal which runs parallel to the mainland between California and Winterton, approximately 1km offshore;
- Holm Channel; Barley Pickle (a deep channel) and Caister Road (also a channel).

Figure 1   Map of study area showing 1992 bathymetry data

The dates of the charts used were 1846, 1864, 1875, 1886, 1896, 1905, 1916, 1922, 1934, 1946, 1954, 1962, 1974, 1982, 1987 and 1992. Maps and charts earlier than 1800 are not usually reliable, in that they are rarely based on an adequate ground-triangulation control, and often were simply sketch surveys, Carr (1962). The data were digitised, reduced to a common projection and datum (OSGB38 Ordnance Survey National Grid) and georeferenced into a Geographical Information System (GIS). The National Grid co-ordinates of the Southwest and Northeast corners of the study area are 650000E, 284000N and 663000E, 319000N respectively. The reduced data were then used to create a TIN (Triangular Irregular Network) model for each survey date. The TIN models were then used to interpolate to a set of digital terrain models, (DTMs), covering the area of the sandbanks at a resolution of 100m. The data on which the charts were drawn had a mean density of 30 points per km^2, corresponding to a mean resolution of approximately 180m.

SANDBANK MOVEMENTS - 1846 TO 1992

To evaluate the general movements of the sandbanks, a visual comparison of the DTM's was conducted using the GIS, see Figure 2.

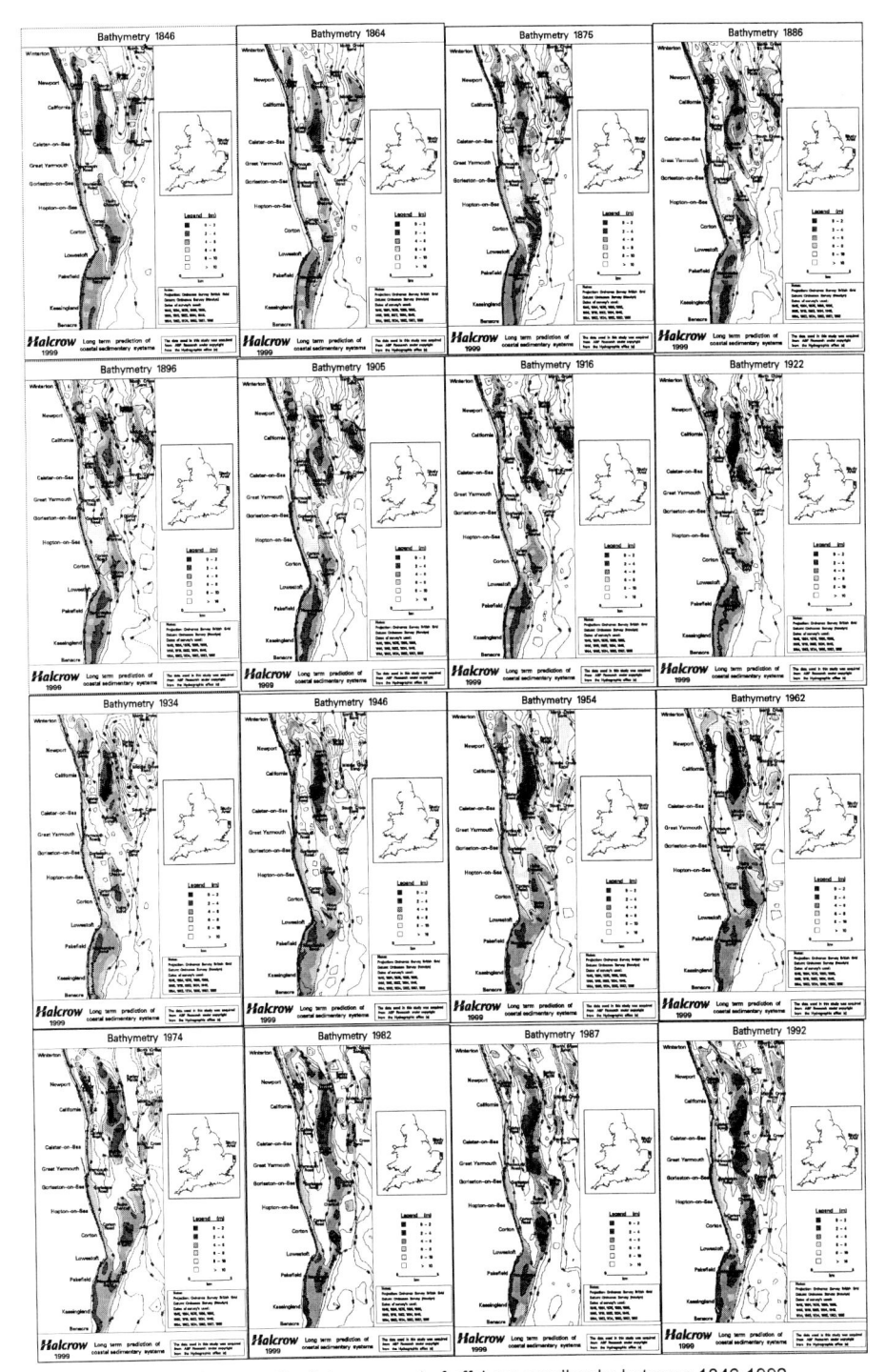

Figure 2 Spatial movement of offshore sandbanks between 1846-1992

As shown in Figure 2, there was a distinct period of erosion between 1846 and 1875, which resulted in an average increase of approximately 1m in the sea bed depth. This change was particularly evident at Holm Sand, where there was a reduction in the height and width of the sandbank, and between Holm Sand and Middle Scroby Sands, where Holm Channel decreased in depth. During this period, there was also some development of the outer Cross sandbanks within the study area.

Between 1875 and 1922, there was further development of the outer sandbanks, especially at the Middle Cross Sands which increased greatly in size. In addition, there was overall shortening and broadening of the main body of Scroby Sands. In the south of the region, there was erosion of the bar off Lowestoft and the formation of a channel across Holm Sand.

From 1922 to 1974, there was accretion across much of the study area resulting in an average overall decrease of 1.4m in seabed depth. The pattern of accretion is clearly visible after 1934, where there was increased development of many of the major sandbanks (e.g. Newcombe Sand, Holm Sand and the Scroby Sands) and disappearance of the channel off Lowestoft. Although the accretion is dominant, there were areas of erosion, notably at the outer South, Middle and North Cross sands. The reduction in size of these features probably relates to entrapment and increased movement of material to the inner sandbanks.

Between 1974 and 1982, the Holm Channel between the main inner sandbanks was closed through increased accretion, creating a continuous bar aligned parallel to the coastline. This sandbank configuration, which is approximately 5km from the shoreline, is still evident today. In addition to the infilling of the channel, there was enlargement of the Middle Scroby Sands and Holm Sand; and enlargement and separation of Corton Sand from the main sandbank system. In addition, the outer Middle and North Cross Sands increased in size reflecting further north-east movement of eroded sediments.

METHODOLOGY

To assess further the movement and variability of the sandbanks over time, mean, range, standard deviation and coefficient of variance maps were produced for the sixteen digital terrain models described earlier.

The mean bathymetry gives an impression of what the sandbanks would look like if all temporal variation were removed. Although this is important, it does not provide any information concerning the magnitude or temporal pattern of movement over time. The range, standard deviation and coefficient of variation can be used to provide a measure of these changes.

The range of bathymetry (difference between maximum-minimum) highlights the absolute changes in bathymetry. However, the range tells us little about the spread of variation over time. This is an important issue, as a large range could be associated with change between two consecutive charts (e.g. large storm event) and is then followed by very stable behavior.

The standard deviation (the square root of the variance) and coefficient of variation (variance divided by the mean) of bathymetry provide a measure of the spread of variation with respect to the mean value over time. The standard deviation is the spread of bathymetry values about the mean, whereas the coefficient of variation is a measure of variability relative to the mean.

Thus for two locations with the same variance in bathymetry levels but different mean depths, the coefficient of variation will be smaller for the location with the deeper mean bathymetry. The coefficient of variation tends to emphasise variability of shallow water (i.e. high bathymetry) such as the sandbanks and beaches.

To supplement these relatively standard statistical measures further GIS analysis has concentrated on assessing the volume of material within the sandbanks. Previous research had estimated the volume of material in the sandbanks to be in excess of 10^9 m^3, Clayton et al (1983). Here, detailed volumetric calculations have been conducted to highlight the magnitude of change over time and also to uncover any temporal pattern of erosion and accretion cycles, which may operate within the sandbank system. The possible implications of these changes upon navigation and the protection of the adjacent coastline area are discussed later.

In addition, we have used Empirical Orthogonal Function (EOF) analysis to investigate the long-term variations which can be resolved by the sequential sampling rate of approximately one decade. EOF methods were developed by meteorologists to analyse time series data, Lorenz (1953). They have since been used to analyse beach behaviour over the period of months or a few years, Winant et al (1975). Here, EOF methods have been developed to analyse the long-term changes in the sandbank configuration.

RESULTS

As shown by the mean bathymetry map (see Figure 3), the main central Scroby Sands, Holm Sands and Newcombe Sands have remained largely stable over the period studied. In addition, it is interesting to note that the pattern shown in Figure 3, mirrors very closely the current configuration of sandbanks, see Figure 1.

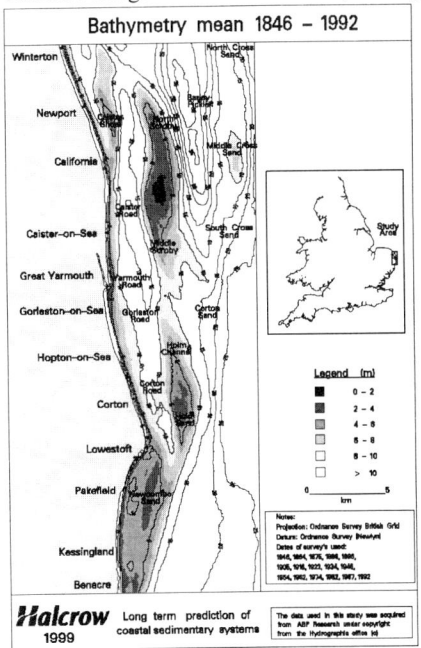

Figure 3   Bathymetry mean 1846-1992

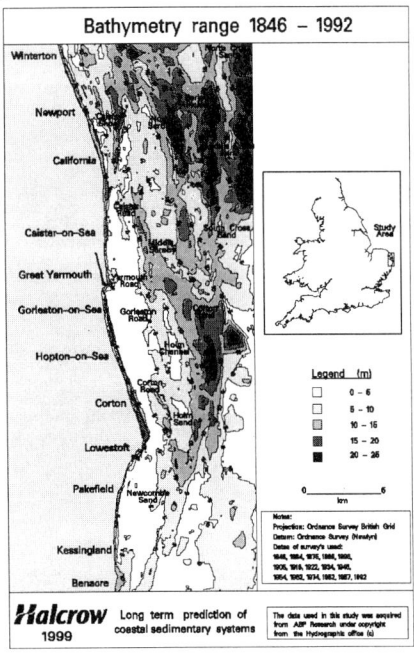

Figure 4   Bathymetry range 1846-1992

In contrast, the only outer sand, which appears to have maintained its general form over time, is the Middle Cross Sands. The variability of the other outer sandbanks tends to suggest that they are more much sensitive to changing patterns of sediment movement.

In the study area, the locations of greatest bathymetric range are situated offshore of Newport and California at Cross Sands and offshore of Hopton at Corton Sand, see Figure 4. In these locations, there has been over 20m change in the period studied. The largest of these changes has occurred on the north-east shelf of Scroby Sands before the drop into Barley Pickle, on the north west flank of Barley Pickle before the rise onto the Cross Sands and on the east shelf of the Middle and North Cross Sands.

The standard deviation and variation maps show that the immediate nearshore area and south of Lowestoft has been the most stable and slow changing, with the main inner sandbanks (e.g. Scroby Sands and Holm Sands) being relatively stable but with greater variability on their outer edges, see Figures 5 and 6. In contrast, there has been considerable variability in the north-east of the study area (e.g. Caister Shoal, North Scroby and Middle Cross Sand) and across Corton Sand offshore of Hopton-on-the-Sea. This pattern re-emphasis that although there has been general growth and movement of these sandbanks in a north-east direction, they remain highly sensitive to changing patterns of sediment movement.

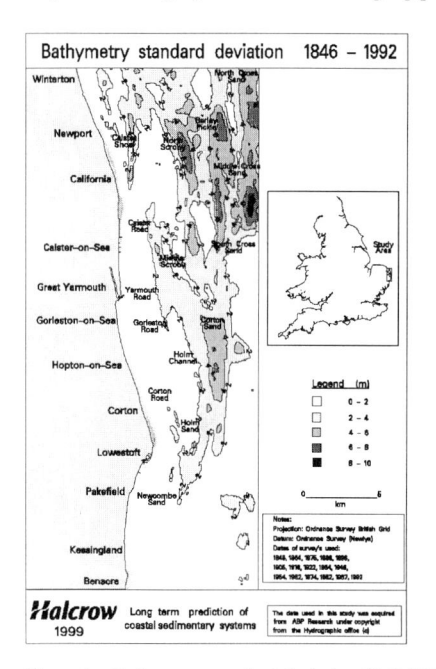

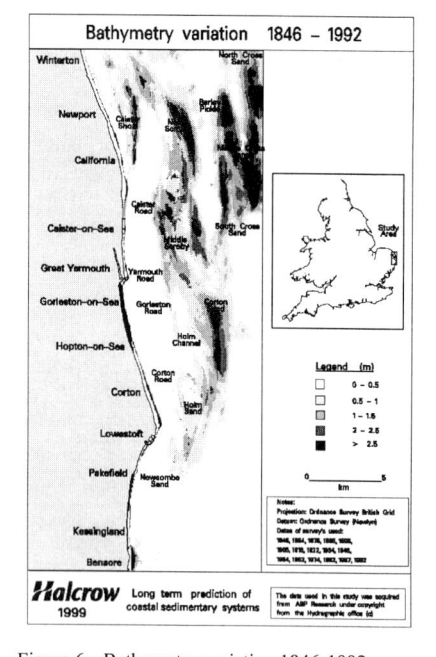

Figure 5   Bathymetry standard deviation 1846-1992

Figure 6   Bathymetry variation 1846-1992

Further GIS analysis has also been used to assess the changes in the average depth of the seabed and the volume of material within the sandbank system, see Figure 7. This analysis has shown that between 1846 and 1875, there was a distinct period of erosion resulting in an average increase of approximately 1m in the sea bed depth. After this phase, there were two smaller cycles of erosion and accretion between 1875 and 1934, followed by a period of accretion to 1974. After this lengthy period of accretion, there was a phase of erosion between 1974 and 1987, and a return to accretion in the last period assessed 1987-1992. Overall, the total volume of sediment within the sandbanks in 1992 has been estimated at between 1.1 and 3.1 x 10^9 m^3. The lower limit is the volume between 0 and 20m below OS datum, while the upper limit is the total volume between 0 and 30m below OS datum. The only previous estimate of the volume of the estimate was the more than 1 x 10^9 m^3 proposed by Clayton et al (1983).

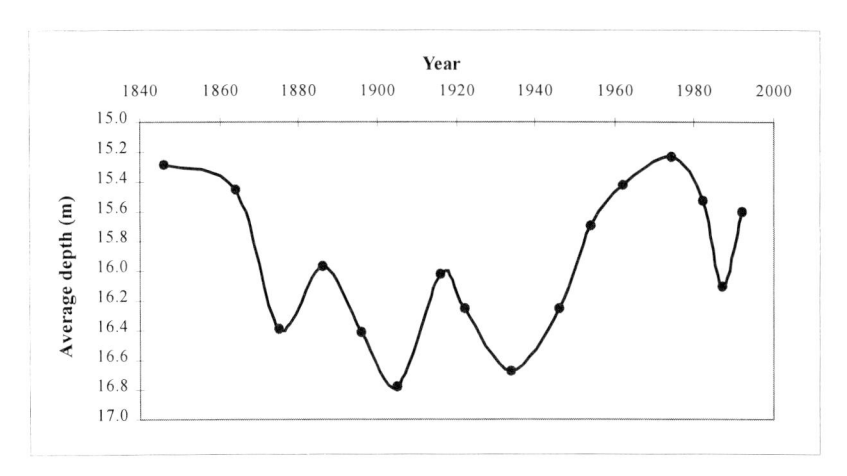

Figure 7   Average depth of seabed in study area between 1846-1992

Table 1 summarises the results of the EOF analysis. It demonstrates that over 97% of the mean square of the data is contained in the first function and that more than 99% of the mean square of the data is captured by only 5 functions. The first eigenfunction corresponds to the mean bathymetry over the period, with the subsequent eigenfunctions representing the variation about the mean. In this case the 2nd to 6th eigenfunctions together account for over 72% of the variance about the mean. This is a large proportion of the variance for analyses of this kind and suggests that the EOF analysis can describe the variability in the sandbank data very efficiently.

Table 1. EOF results

Order	Normalised eigenvalue	% mean square (cumulative)	% variance	% variance (cumulative)
1	0.9735	97.4	-	
2	0.0085	98.2	32.0	32.0
3	0.0033	98.5	12.5	44.5
4	0.0029	98.8	10.9	55.4
5	0.0023	99.1	8.6	64.0
6	0.0022	99.3	8.1	72.1

The EOF analysis expresses the data as a sum of products of functions depending on, individually, space and time. It thus provides a means of separating variations in space and time, which have distinct frequencies and spatial scales. Figure 8 shows the graphs of the first four temporal eigenfunctions, which describe the variation in time of the corresponding spatial functions. There are several features of note:

84   COASTAL MANAGEMENT

- The first function is almost constant, as expected because it corresponds to the mean;
- The second function shows a slow variation from negative to positive values over the period and is, arguably, indicative of a recurrent variation with a period of the order of 200 years;
- The third function shows a sharp drop from positive to negative values from 1840 to 1880 and smaller variations subsequently. This is indicative of an isolated switch in behaviour followed by relatively stable behaviour;
- The fourth function shows a variation in time which is strongly suggestive of periodic behaviour with a period of 100-120 years.

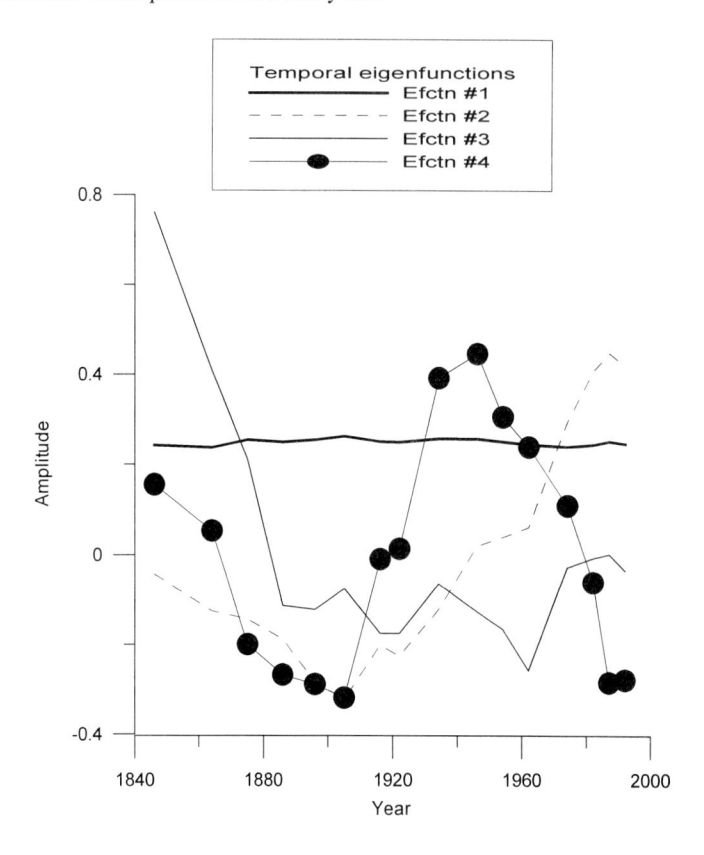

Figure 8. Temporal Eigenfunctions

Figure 9 is a contour plot of the second spatial eigenfunction, which exhibits strong spatial structure. Minima are found on the seaward flanks North Scroby Sand, South Cross Sand and North Cross Sand, and around Holm Channel. Maxima occur in the southern end of Barley Pickle, and the landward flank of North Cross Sand. Variations associated with this eigenfunction will be in the same sense wherever the function is the same sign. Thus, for example, variations in seabed levels at Holm Channel and North Scroby will be highly correlated. This correlation will not necessarily be apparent when the contribution of the other eigenfunctions is also considered at the same time.

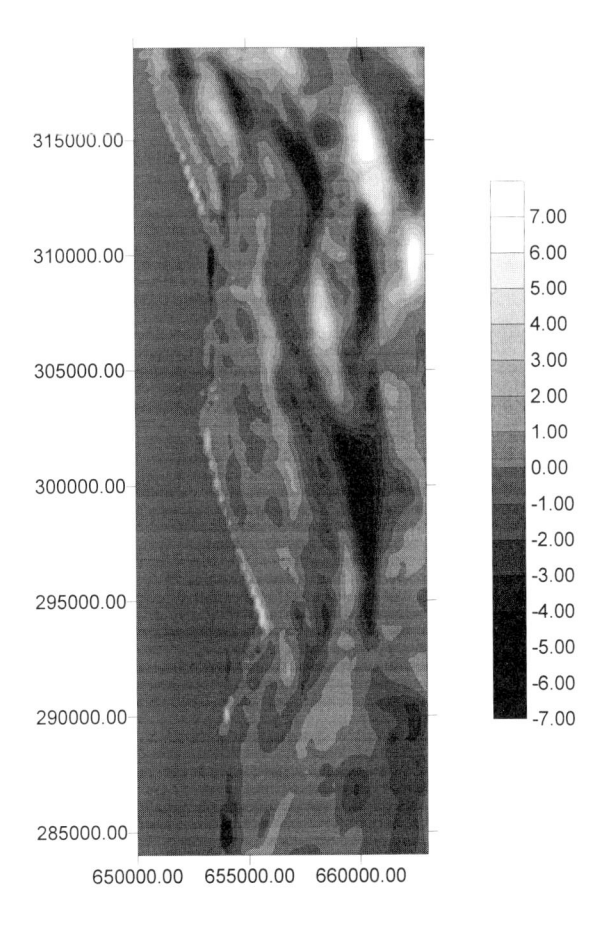

Figure 9. Spatial eigenfunction #2

DISCUSSION & CONCLUSIONS

The analyses have identified long-term trends in volume and areal extent, as well as evidence for oscillatory behaviour within the sandbank system. Evidence of variations, which had several distinct cycles within the period covered by the study, would increase confidence in the predictability of the variations. However, the presence of trends and recurrent behaviour provides some scope for predicting the future development of the sandbanks. In this regard it could be noted that the period of 'oscillations' identified in the eigenfunctions is greater than the time horizon currently adopted in most SMPs. This raises the possibility that long-term recurrent behaviour is treated as an underlying trend on the planning timescale. The consequences could be a planned situation that is unsustainable when it becomes apparent that the 'trend' is actually a long-term 'oscillation'.

REFERENCES

Carr, A.P., 1962. Cartographic record and historical accuracy, Geography, 47 (2), p135-144.

Clayton, K.M., McCave, I.N. & Vincent, C.E., 1983. The establishment of a sand budget for the East Anglian coast and its implications for coast stability, Shoreline Protection Conference, Thomas Telford, p91-96.

Halcrow, 1991. The Future of Shoreline Management, Conference Papers, Publ. National Rivers Authority Anglian Region & Sir William Halcrow & Partners Ltd., pp116.

Li, B. & Reeve, D.E., 1999. Analysis of long-term changes in nearshore morphology, to be presented at IAHR Symposium on River, Coastal and Estuarine Morphodynamics, September, 6th-10th 1999, Genova, Italy

Lorenz, E.N., 1956. Empirical orthogonal functions and statistical weather prediction, Rep. 1, 49pp, Dept of Meteorol., Mass. Institute Technol., Cambridge.

Raper, J., Livingstone, D., Bristow, C. & McCarthy, T, (1997) Constructing a geomorphological database of coastal change using GIS. CoastGIS '97. Second International Symposium on GIS and Computer mapping of Coastal Zone Management, Aberdeen, August 1997.

Reeve, D.E., 1995. Stochastic Methods for Maritime Resource Stewardship, PIANC Bulletin No88, p5-21.

Robinson, A.H.W., 1960. Ebb-flood channel systems in sandy bays and estuaries, Geography, 45, p183-199.

Sims, P.C., Weaver, R.E. & Redfern, H.M. (1995) Assessing coastline change: A GIS model for Dawlish Warren, Devon, UK. CoastGIS '95: International symposium on GIS and computer mapping for coastal zone management, Cork, Ireland, February. 1995.

Winant, C.D., Inman, D.L. & Nordstrom, C.E., 1975. Description of seasonal beach changes using empirical eigenfunctions, J. Geophys. Res., 80(15), p1979-1986.

ACKNOWLEDGEMENTS

This work was funded under contract by the Ministry of Agriculture, Fisheries and Food, Contract No. FD1008 to Halcrow Group Ltd.

Predicting Long-Term Coastal Morphology

H N SOUTHGATE, A H BRAMPTON, and B LOPEZ DE SAN ROMAN
HR Wallingford Ltd, Wallingford, Oxon, OX10 8BA, UK

ABSTRACT
This paper reviews some of the ways that coastal managers can predict shoreline evolution, concentrating on long-term (yearly to decadal) variations in beach and nearshore morphology. It contains a summary of presently available models for coastal morphological predictions on these timescales, and explains why traditional deterministic predictions are often scientifically unsound. Various sources of error are considered, and it is concluded that probabilistic predictions are, in many cases, the best option. Careful considerations of sources of error and methodologies for using models to obtain probabilistic predictions are regarded as necessary in design calculations. Much of the work on which this paper is based has been carried out as part of the MAFF CAMELOT programme (Coastal Area Modelling for Engineering in the LOng Term; Southgate, 1998).

INTRODUCTION
The prediction of the future morphology of beaches is not only an important element in deciding whether to undertake coastal defence, but also as part of the design of defences themselves. However, there are a number of reasons why it is not possible to predict shoreline changes with any degree of certainty in the medium to long-term, i.e. over periods of up to years and decades.

- First, the weather conditions that create waves and influence currents and water levels are not predictable for more than a few days in advance. For this reason, we are forced to adopt a "climatic" view of these hydrodynamic conditions, i.e. adopting a probabilistic approach. Our knowledge of the present and recent past weather conditions, however, is often incomplete.
- There is a lack of long-term field data with high resolution in space and time, for calibrating and verifying models of long-term coastal morphology.
- The basic properties of coastal morphology suggest that morphological changes can have a chaotic type of behaviour. If so, this would mean that accurate predictions would be impossible beyond a certain time into the future, however good our predictions of waves, currents and sediment transport become.
- Another fundamental limitation to long term morphological predictions is that we do not know (and cannot know) the future *sequence* of meteorological events, even if the climate is accurately known. The effects on beach profiles and planshapes of the same wave conditions occurring in a different order can be important.

For these reasons, in many typical engineering applications, the traditional "deterministic" approach involving a single model run into the future has serious drawbacks and cannot be justified scientifically. In these circumstances, probabilistic predictions are the best option.

Although there is some scope for developing models that directly deal with probabilistic quantities, much can be done with existing deterministic models by adopting different methodologies to produce probabilistic predictions. These methodologies involve:

- Multiple model runs to build up probability distributions of the morphological variables of interest
- Quantifying sources of error, and determining their effect on the final predictions.
- Validation of models, both as general tools and for specific sites or applications

PURPOSES FOR MAKING PREDICTIONS OF COASTAL MORPHOLOGY

At the outset of any project in which coastal morphological models are envisaged, it is important to establish the *type* of morphological prediction required. Some examples of these prediction types are:

- Prediction of the initial rates of sedimentation and erosion
- Prediction of morphology at a specified future time
- Prediction of morphology over a future timespan
- Prediction of time taken for a specified morphological state to be reached with a specified probability (e.g. for a channel to infill to a certain level, or the shoreline to erode to a certain line)
- Predictions involving different timescales (e.g. long term trends plus short term variability)

In deciding which type of morphological prediction to use, it is necessary to consider:

- The requirements of, and constraints on, the project
- The capabilities of available models that could be used in the project

Another requirement at the outset of a project is to decide the scenarios to be investigated by morphological modelling. These scenarios usually comprise one or more of the following:

- Do-nothing option
- Different natural scenarios (e.g. investigate different scenarios of future sea level rise or changes in wave climate)
- Different human interventions (e.g. introduction of hard or soft defences)

TYPES OF NUMERICAL MODELS

Numerical models of coastal/ nearshore morphology are categorised both by considering how they deal with morphological changes in *time*, and how they deal with morphological changes in *space*. Most models belong to one type in both the "time" and "space" categories.

Categorisation of coastal morphological models by "time"

The way models deal with changes in time can be described by introducing three classes termed "equilibrium", "initial sedimentation and erosion (ISE)" and "sequential" models. These are described below.

- *Equilibrium Models* are designed to produce a final equilibrium morphological shape for any prescribed input conditions. The final morphology is independent of the initial state and in some cases no information is given on the rates of sediment transport or on time taken for the changes in morphology to occur. These models assume that the morphology does evolve to an equilibrium state for the considered forcing conditions. Generally, this will not always be the case, and some evidence (from data and/or theory) should be put forward to support this assumption.

- *Initial Sedimentation and Erosion (ISE) Models* determine the *initial* rates of erosion and accretion of the seabed. There is no calculation of how the seabed evolves in time, although for short future periods, this can be inferred from the calculated initial rates of change of seabed levels. Further into the future, of course, these rates of change themselves will change, so this simple extrapolation of the initial values is not possible. Often these models are the same as the sequential models described below, in which the morphodynamic updating facility is not used.

- *Sequential Models* involve calculating the rates of seabed erosion and accretion, and then updating the morphology of the seabed/ beaches over a certain time into the future (the "timestep"). The size of this timestep has to be determined carefully, for example, it must be small enough to resolve important changes in the wave and tidal forcing conditions and to prevent numerical instabilities, but not so small that the computational effort is excessive to reach the desired simulation time. After one timestep, a new forcing condition is specified, and the process of calculating transport rates and morphology changes is repeated for this new condition. These models may be run for an indefinite period, until perhaps the morphology changes become very slow and insignificant (an equilibrium configuration), or for a specified time, for example to calculate sedimentation in a dredged area during a single tidal cycle. These models assume that the variation of forcing conditions over time is sufficiently gradual that time derivatives, e.g. accelerations in flows, are negligible. Hence considering a sequence of "steady state" situations, for example at different stages during a tidal cycle, represents the effects of time variations. Such models can provide results on the statistical variations in morphology as well as a long-term trend. However, the computational effort can be high, and errors usually accumulate with increasing numbers of timesteps.

Categorisation of coastal morphological models by "space"

There are also three main types of numerical model for different coastal spatial configurations.

- *Coastal Planshape Models (Longshore Models).* These models are one-dimensional, with the X-axis established approximately parallel to the coastline. The beach morphology is represented by a single contour and this, together with other parameters (e.g. wave conditions, currents, sediment transport rates) are functions of only X and time t. Predictions of changes in the beach and nearshore seabed plan shape are produced. The beach profile is assumed to be constant with time, although some variation in profile along the shoreline is sometimes allowed.
- *Coastal Profile Models (Cross-shore Models).* These models are also one-dimensional, with the X-axis established approximately perpendicular to the coastline. They predict the changing shape of the beach/ nearshore seabed profile, but usually there is no or only limited representation of longshore variations in sediment transport or morphology. As well as predicting beach profile changes, and the movement of sediment perpendicular to the contours, some of these models will represent changes in wave conditions and currents (wave- or tidal-driven) along the shore normal.
- *Coastal Area Models.* Where currents and sediment transport pathways are not shore-parallel or shore-normal but have significant components in two dimensions, the simplifications made in the above two types of model are unlikely to produce accurate

results. The traditional approach to such situations has been the physical model, and this still may be the best option for small areas where only a limited range of wave/ current conditions need to be tested. However, a number of new "Coastal Area" computational models have been developed over the last ten years or so, and these are now beginning to be used for practical coastal management problems.

A two-axis system is established, usually with one axis parallel to the shoreline (**X**), the other perpendicular to it (**Y**). The hydrodynamic motions (waves and currents) are calculated as functions of **X**, **Y** and time **t**. Coastal area models are of the sequential type, as described above. A digital representation of the initial morphology of the beach and/or nearshore is required together with boundary conditions for the wave and tidal forcing.

SOURCES OF UNCERTAINTY IN MORPHOLOGICAL PREDICTIONS

In any design study, it is important to understand how significant are all the various sources of uncertainty. Furthermore, it is important to distinguish between sources of *quantifiable* uncertainty, and *unquantifiable* uncertainty. The former refers to sources of uncertainty about which calculations can be made of the spread of possible outcomes of coastal morphology, usually in the form of probabilities of occurrence. In the latter case, these calculations are, for whatever reason, impossible.

Unquantifiable Uncertainty

Examples of sources of unquantifiable uncertainty are:

- Lack of relevant data
- Lack of knowledge of relevant physical processes
- Inability to incorporate relevant processes in models
- Required parameter ranges of processes not covered in models
- Lack of methodological knowledge for models

Uncertainties in coastal morphological prediction can differ greatly for different morphological features, nearshore zones and timescales. Obviously, in any design study it is important to reduce sources of unquantifiable uncertainty as much as possible. For example:

- Existing sources of input data can be supplemented by additional measurements or data from larger-scale models
- Lack of scientific or methodological knowledge cannot be usually addressed as part of a design study, unless the knowledge is highly specific to the study. In the longer term, these sources of uncertainty are the subject of strategic research.

Quantifiable Uncertainty

Sources of quantifiable uncertainty are those for which calculations can be made about the spread of possible predictions of future coastal morphology. Recent research has been aimed at devising methodologies of using morphological models to do these types of calculation. Examples of these sources of quantifiable uncertainty are:

- Random errors in input data
- Systematic errors in input data
- Errors in calibrated parameters or internal model parameter settings
- Interaction of errors in model (especially if non-linear)

- Accumulation of result errors through simulation time
- Lack of knowledge of future wave sequences (and other essentially random forcing factors such as wind)
- Chaotic or other inherently unpredictable types of long term behaviour characteristic of non-linear systems like coastal morphology.

Several techniques have been developed to quantify the effects on coastal morphology of the various sources of quantifiable uncertainty. Two common themes to all these techniques are the use of multiple model runs, and the determination of probabilistic results.

- Random and systematic errors in input data can be addressed by multiple model runs with varying values of input parameters within a specified range (Vrijling and Meijer, 1992)
- Errors in calibrated parameters or internal model parameter settings can be treated in a similar way
- The effects of wave chronology can be assessed by multiple model runs using reordered versions of the wave sequence as input (Southgate, 1995)
- The limits of predictability in a chaotic regime can (in principle) be determined by ensemble modelling, in which multiple model runs are performed with slightly different values of the parameters that determine the initial state. However, at the present time, models have not been developed for the purpose of studying possible chaotic behaviour. Until such models are developed, this source of uncertainty has to be regarded as unquantifiable.

VALIDATION OF MORPHOLOGICAL MODELS

Definition
Model validation refers to testing of models against data, and, if required, subsequent amendments to the model to improve predictions.

However, 'validation' has a number of more precise meanings in different contexts:

- *Software Industry*. Validation refers to well-defined procedures for model development, testing and documentation.
- *Academic Research*. Validation mainly relates to the accuracy of model predictions of physical parameters in relation to detailed data sets.
- *Engineering Consultancy*. Validation refers to the usefulness of model predictions in real or simulated engineering design problems.

'Validation' is used here in the academic and engineering senses. The word 'verification' is also used, often interchangeably with 'validation'.

Validation also relates to the intended scope of the model.
- *Global validation*. Most morphodynamic models are developed as generic tools, i.e. they are designed for a range of coastal geometries, sediment types, hydrodynamic forcing etc. A thorough validation exercise therefore would need to cover the full range of intended applications of the model.
- *Local validation*. When a morphological model is used for a particular project or site, the required range of validity is usually much less than full range for which the model is

designed. Validation for this particular use of the model therefore need focus only on the parameter ranges that pertain to that use.

Aim

The aim of a validation exercise is to compare model predictions with data of sufficient quantity and quality, covering the desired range of application of the model, so that reliable quantitative conclusions can be made about the model's performance as a predictive tool.

However, available field data is generally much less than would be required for a thorough global validation. In addition to lack of data, validation of a morphological model is made more difficult by the following factors:

- *Inherent variability of sediment processes.* Much less accuracy can be achieved in predictions of parameters involving sediment processes than is possible in purely hydrodynamic comparisons (waves, currents, water levels)
- *Non-linear interaction between hydrodynamic and morphological parameters.* These interactions may sometimes lead to strong and surprising morphological behaviour. A denser coverage of data through the validation range would therefore generally be needed compared to a linear system.
- *Calculations through time.* As calculations proceed through time, model predictions generally become more inaccurate because of cumulative errors in the calculations. In some cases, models will be used to make probabilistic rather than deterministic predictions.

Generally, morphological models are sufficiently complex, and data sufficiently scarce, that model validation cannot be done to the level at which firm quantitative conclusions can be made. Accordingly, validation involves comparing model predictions with data that is available, and the model performance has to a significant extent to be subjectively assessed by people with relevant expertise. Because of this important subjective role, weaker terms such as 'comparison' or 'evaluation' may describe the validation process more appropriately.

Procedure

The validation procedure involves the following steps. At this stage, validation data will have been identified, assembled, and if necessary processed.

- Define the input parameters for the model and their ranges of values for which the model is designed to be valid.
- Decide whether 'global' or 'local' validation is to be attempted
- Decide which model output parameters are to be compared with data, and in what form (e.g. deterministic or probabilistic)
- Perform model runs and store output. In some cases, multiple model runs may be needed, such as for probabilistic comparisons.
- Compare model output with data. This can be done in several ways (we consider here that the output consists of bed levels over a coastal area after a certain simulation time, but other types of deterministic output can be considered in a similar manner). This can be done in several ways, namely:
  1. *Visual inspection.* An experienced modeller should always visually inspect at least some of the validation test results. The human eye is adept at picking out patterns of behaviour. In some cases, gross discrepancies may be visible, which would indicate

that corrections or development of the model are needed, without recourse to further quantitative comparisons.
2. *Analysis of both model results and data, and comparisons of analysed results.* The analysis can take the form of, for example, spatial averages, variances, spectral analysis or principal component analysis. In each case the aim is to isolate particular features of the data and model results for comparison.
3. *Quantitative skill assessment.* 'Skill' is used in a technical sense as a quantitative measure of the accuracy of model predictions with data, compared with the accuracy of a simple default case (often this is just the initial morphological state) using the same data. Various methods can be used to calculate the 'skill score', such as the 'Brier score'. The technique can be applied to spatial subsets of the model area in order to focus on particular morphological features. Skill assessment has not been widely used in coastal morphological modelling but is a common technique in meteorological forecasting (Murphy and Epstein, 1989).

- Model validation can be carried out for probabilistic results from morphological models. In these cases, multiple model runs are performed, and the results processed to form probability of exceedance plots of seabed levels. These can then be compared with data to see if the data points lie within the model probability bands. Obviously, the wider the probability bands, the greater the chance of agreement with data, but wider probability bands also reduce the predictive value of the model. Probabilistic validation has generally been little used in coastal morphology, and there is the opportunity to learn from procedures in other disciplines such as weather forecasting.

SEQUENCING OF WAVE CONDITIONS (WAVE CHRONOLOGY)
Wave chronology refers to the effects on coastal morphology of different sequences of input data, in which the actual sequence is not known, but the probability distribution of forcing conditions can be determined reasonably accurately. Usually, the forcing conditions refer to input wave conditions over timescales greater than a few days, but they can refer to other input such as wind conditions or water level surge, whose actual sequence is largely unpredictable. On the other hand, input sequences of tidal phenomena (water levels and currents) can often be predicted accurately over time periods of up to decades, so chronology is not an issue.

Scientifically, it is the *non-linear* response of coastal morphology to wave input that makes wave chronology potentially important. If the response were linear, it would be independent of the sequencing of wave conditions, and would depend only on the overall statistics of the whole wave sequence.

One type of question that can be asked about chronology concerns the sequencing of storms. For example, what are the effects on morphology if storms occur at the end of a wave sequence compared with at the start or if they are evenly spread throughout the sequence? The chronology of more moderate wave conditions may also have a significant role.

The following steps describe a methodology of assessing the effects of wave chronology on coastal morphology.

Establish the General Purpose
One can study wave chronology for two general purposes:
- To assess the general effects of wave chronology at a site using past data
- To make future predictions of morphology that take account of wave chronology effects

Derive a Sequence of Wave Data

The starting point for a study of wave chronology is a sequence of wave conditions with the following properties:

- It pertains to a location at the seaward boundary of the morphological models to be used
- It contains the wave parameters of interest (usually wave height, period, direction and water level)
- The duration of the wave data covers the past or future timespan that one wishes to study
- The wave data is at equally spaced time intervals

A suitable wave data sequence can be derived in two ways:

- Directly from measurements or other models
- Synthesised from probability distributions of wave parameters

Splitting and Recombining the Wave Sequence

The wave sequence should be split into a number of segments. The choice of these segment boundaries should respect the following constraints as far as possible.

- At least four segments should be made
- Individual meteorological events (storms etc) should be included in the same segment
- The length of segments should be greater than the period of predictability of meteorological events, and the time of travel of distantly generated waves with significant levels of energy (in both cases, typically a few days)
- Segment boundaries should be at similar low wave height values, to avoid discontinuities when the segments are rearranged
- Cyclic features in the data should be preserved. This most commonly applies to the annual cycle in data spanning several years. In this case, segments should be at yearly intervals or simple multiples or fractions of a year.

Segments should then be recombined in several different ways to form wave sequences of the same length as the original. Each reordered wave sequence will have the same time-independent statistical properties as the original (e.g. average and variance). This reordering of segments should be done for each independent relevant wave parameter (usually, wave height, period, direction and water level).

Running the Model and Output

The model should be run for each reordered wave sequence. After each run, the final bathymetry should be stored. Depending on the type of predictions required, other model output should be stored, such as bathymetric values at intermediate times. More information about this additional output can be found in Southgate (1995).

Interpreting the Results

These model results can be interpreted in terms of the effects of chronology on the future bathymetric levels. This interpretation is quantitative and probabilistic. It follows a definite procedure that can be automated as a post-processing exercise.

This type of processing gives results as envelopes of bathymetry (covering the spatial domain) corresponding to specified probabilities of exceedance. Typical values of these probabilities would be:

- 50% (envelope of mean levels)
- 32 % and 68% (envelope of mean level ± standard deviation)
- 5% and 95% (envelope of mean level ± 2 x standard deviation)
- Envelopes of maximum and minimum levels

Figure 1 shows an example of the envelope of maximum sediment build-up around a groyne, taking account of chronology effects using the above procedures. The general principles of wave chronology are outlined in Southgate and Capobianco (1997). Applications to planshape modelling are given in Lopez de San Roman and Southgate (1998) and to profile modelling in Southgate (1995).

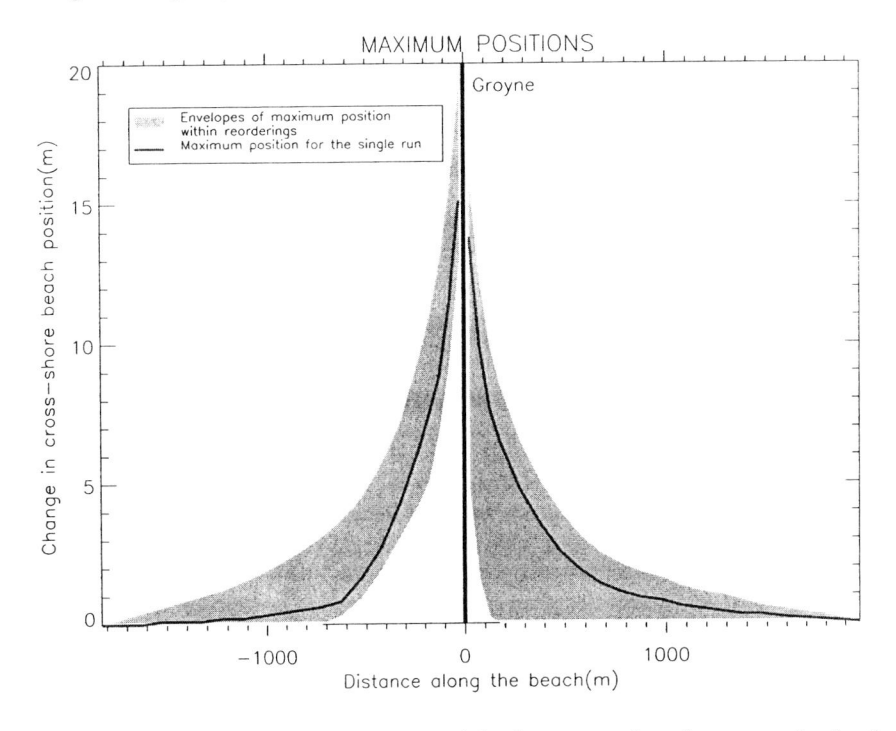

Figure 1 Envelope of changes in a straight beach following construction of a groyne, showing the effects of wave chronology using a planshape model. Duration = 5 years. Number of reorderings of the wave sequence = 40. For each wave sequence, the maximum (most seaward) position of the beach during each of the 40 5-year sequences is recorded. One can then draw 40 separate envelopes of maximum beach position. The shaded area shows the region covered by these 40 envelopes. The continuous line shows the envelope of maximum beach positions from one of the wave sequences. Note that the wave conditions were distributed approximately evenly about the groyne direction, accounting for similar accretion patterns on both sides of the groyne.

CONCLUSIONS

Predictions of how coastal morphology evolves over periods of up to decades are an essential part of many coastal engineering design studies. However, these predictions are usually subject to substantial uncertainties, which become larger the further into the future that predictions are made.

Some types of prediction error may be reduced by improvements to the representation of processes in models and by more morphological data for validation. However, there are also some inherent scientific limitations to the accuracy of morphological predictions. In these cases it is better to adopt probabilistic approaches, which can involve new methodologies for using models. Some of these methodologies have been outlined in this paper.

Probabilistic predictions of coastal morphology can be incorporated into a wider modelling framework involving other engineering design parameters. An example would be to incorporate probabilistic predictions of beach levels with those of extreme waves and water levels. These morphodynamic predictions can also form an element of, or interface with, decision-making software such as Geographical Information Systems.

REFERENCES

Lopez de San Roman Blanco, B and Southgate H N (1998), The effects of wave chronology on beach planshape modelling, HR Report TR 67

Murphy A H and Epstein E S (1989), Skill scores and correlation coefficients in model verification, *Monthly Weather Review, 117*, 572-581

Southgate H N (1995), The effects of wave chronology on medium and long term coastal morphology, *Coastal Engineering, 26, (3/4)*, 251-270

Southgate H N (1998), The CAMELOT programme for predicting long term coastal morphology. An overview from HR Wallingford, *33rd MAFF Conference of River and Coastal Engineers*, Keele University.

Southgate H N and Capobianco M (1997), The role of chronology in long term morphodynamics: Theory, practice and evidence, Proc. Coastal Dynamics '97, Plymouth, UK, 943-952

Vrijling J K and Meijer G J (1992), Probabilistic coastline position computations, *Coastal Engineering, 17*, 1-23

The Contribution of Technology in Supporting Data Exchange in Integrated Coastal Zone Management.

KEIRAN MILLARD, MPhil. MSc. BEng & AMANDA BRADY MSc. BSc.
HR Wallingford, Oxfordshire, UK

INTRODUCTION
Alongside the institutional and funding mechanisms, it is well recognised that an information provision mechanism is the third core component of any integrated coastal zone management (ICZM) plan. Whilst during the early and mid-nineties the focus of debate was heavily on the institutional components of ICZM, recent emphasis has been on the deficiencies in the information provision mechanism. Those responsible for establishing and implementing ICZM plans have found that collating and analysing the data necessary to underpin their decision making is hampered by traditional, sectorially based, data-capture procedures. Over many years, the different activity sectors operating in the coastal zone (e.g. dredging, fisheries, power generation and coastal defence) have evolved specific routes to satisfy their own information requirement. These information provision routes involves specific funding policies, data management policies and technologies for data management.

The telematics community has sought to address this problem by using technology to aid in the sourcing, interoperability, integration and visualisation of data. In many cases the aim is to provide the coastal manager with 'information on demand' by which the coastal manager simply requests to a networked computer what data is required and it is presented in a readily accessible form. The techniques used in these processes include a combination of advanced technologies such as Geographic Information Systems (GIS), Common-Object Request Broker Architectures (CORBA) and the Internet in conjunction with cataloguing and metadata standards. There is no doubt that such technologies can be used to build information systems to support ICZM, but technological issues are only one barrier in information provision.

In this paper the contribution of information technology (IT) in supporting data exchange in ICZM is examined. Using as an example the EC-funded projects THETIS, this paper reviews how successful technology can be in advancing data exchange. This considers the technologies used and the importance of metadata systems.

PROVIDING INFORMATION FOR ICZM
Integrated Coastal Zone Management (ICZM) is a 'holistic' management technique that recognises that the coastal zone contains a finite set of resources that support the numerous activities taking place there. Competition between these sectors has resulted in unsustainable resource use and environmental degradation. Therefore the focus of development in ICZM over recent years has been to set a framework for institutional co-operation.

To support institutional co-operation, ICZM also requires a mechanism for co-ordinated information provision (OECD, 1992). However, the sectoral participants in coastal management have, over many years, evolved specific routes to satisfy their data requirements. These data provision routes encompass specific data collection methods, data processing and storage technologies, and data control policies. These has resulted in a number of barriers that prevent data being effectively translated into information (CIRIA, 1999):

- Data users cannot identify the existence or location of data. Conversely, some users experience data overload and difficulties in identifying relevant or reliable data.
- There is a lack of documented context information describing data, which impedes the transformation of raw data into information that planners, decisions makers, and the public can understand and use. This limits the added value potential of the data.
- Narrow visions in data collection investments and campaigns, with limited, if any, consideration of the value of data to other users. This results in an unnecessarily low cost-benefit ratio of data items.
- Technology and data format barriers impede data producers, brokers and consumers to communicate with each other, resulting in data and computation islands.
- The lack of data quality standards and an insufficient auditing of the quality of data holdings. This may undermine the usefulness of data and its 'trustworthiness'.

In the early-mid 90's, the European Commission stressed the need for developing information technology infrastructures facilitating people and data networking to improve decision making at the public administrator/coastal manager. Recently, the EC has completed a demonstration programme on ICZM and this programme culminated in the DGXI report on the "lessons learnt from the demonstration programme" (EC, 1999). This report stated that there is "a need for a comprehensive approach to pro-active and co-ordinated diffusion of existing information in Europe and that efforts to date in this realm have been insufficient".

Generating usable and dependable information to support ICZM requires integration of varied data on geographical, socio-economic and bio-geophysical parameters. Practically, this data is distributed over various places at local, regional, national and/or international level. Exploitation of such distributed data requires data mining and data transfer mechanism, processing and modelling tools, and geographical data integration tools such as Geographic Information Systems (GIS). The skills required embracing all these tools and mechanisms are beyond those of the coastal manager. Accordingly they need to be seamlessly embedded into a coherent processing infrastructure.

On-going collaborative research at HR Wallingford is examining using information technology systems to provide information appropriate to the level of the coastal manager. These systems typically integrate data from a number of remote data repositories and process the data in response to user requests for information. The concept of this technology system is shown below in Figure 1.

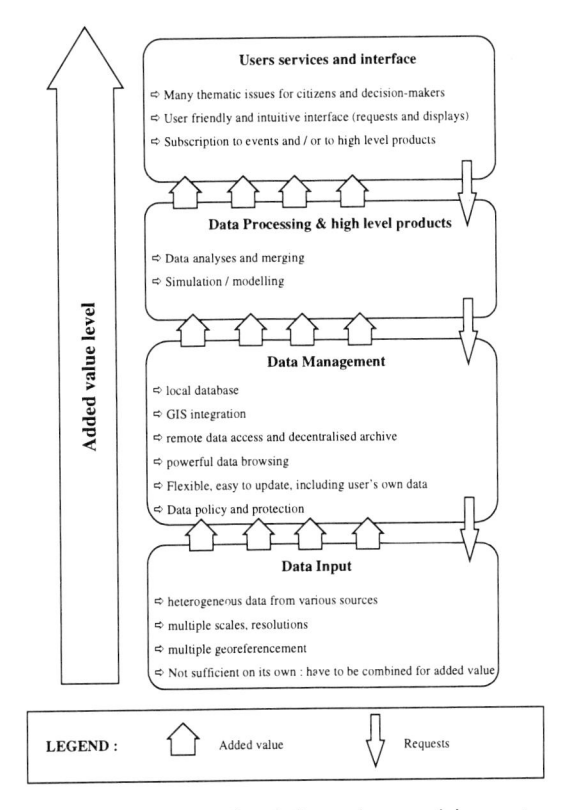

Figure 1. The concept of an information provision system

METADATA AND COASTAL ZONE MANAGEMENT

Metadata

Metadata provides an important cornerstone to any information provision system used in ICZM. Metadata is information about data. It is a description of the data characteristics (e.g. the data format) and processing history of the data (e.g. location and owner). It is used to enable the human user to determine if the data is of use (content metadata) and increasingly used for automated data searching and processing (machine metadata).

Good metadata contains a lot of descriptive terms about the data content, theme, spatial coverage, and time period. These are the topics that are searched on by the user and required to assess fitness for use (i.e. data quality). The information contained in metadata is often no different to the descriptive elements of reports, but simply structured differently to meet the requirements of metadata standards.

Much of the data required for coastal management is now in electronic format and is increasingly exchanged via the Internet. Accordingly, measures to standardise information formats and facilitate interoperability between users are likely to have the greatest impact on data accessibility. Information for ICZM is produced and consumed by many diverse organisations that have distinct data requirements, and so in most practical cases, data exchange will be mediated by metadata.

The more detailed and content-rich the metadata, the easier it is to find and use data. Figure 2 shows some simple metadata that may accompany a data set. When writing metadata it always useful to consider 'what information would help me to use this data more easily'.

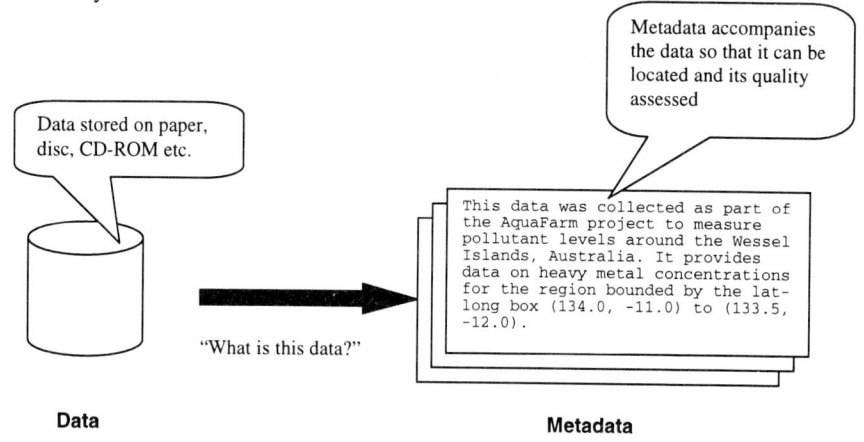

Figure 2. Example Metadata

Metadata Language and Standards

No universally applied metadata standards are in existence. Most organisations use their own procedures, driven by their data sources and user needs. The metadata community has developed a range of standards, but few can be regarded as 'common knowledge'. An example of some standards is given in table 1. This table also includes 'thematic provider standards'. These are not strictly metadata standards, but offer a uniform way of describing data.

Table 2. Examples of global metadata standards (CEO, 1996)

Standard	Description
ANZLIC	Australia/New Zealand geographic metadata standard
CEN TC 278 / CEN TC 287	Geographic metadata standards
CIP	Full interoperability standard for EO data catalogues
EDMED	Thematic provider standard
FGDC	Geographic metadata content standard
GDDD (MEGRIN)	Geographic metadata standard
GILS	Geographic /EO metadata standard
ICES	Oceanographic metadata standard
ISO TC 211	Metadata standard for all Earth related
ROSCOP	Thematic provider standard

For writing metadata, two international standards can be recognised: SGML (Standard Generalised Markup Language) and XML (eXentsible Markup Language). SGML is an International Organisation for Standardisation (ISO) standard that is used to define the structure of electronic text files or documents. It is concerned primarily with structure and not the content of the document. HTML (hypertext mark-up language) is a well-known sub-set of SGML. XML is designed to bring the benefits for SGML to the web, namely the ability to handle large and complex documents and the ability to define your own class of documents with their own unique structure. It is fully compliant with the ISO SGML standard.

The elements, order and structure of an SGML file are defined in another document file called a DTD (Document Type Definition). For example, there is a DTD that defines HTML, likewise there is a DTD that defines a standard set of elements and standard structure for files of such as ANZLIC metadata using SGML (see Figure 3).

```
<descript>
    <abstract>
        <p>This is the abstract.  This data was collected as part of the
        AquaFarm project to measure the pollutaqnt lelvels around the Wessel
        Islands, Australia
        </p>
    </abstract>
    <theme>
        <keyword qualifier="monitoring">WATER</keyword>
        <keyword>AQUACULTURE</keyword>
    </theme>
    <spdom>
        <place>
            <dsgpolyo>
                <long>134.0</long>
                <lat>-11.0</lat>
                <long>133.5</long>
                <lat>-11.0</lat>
                <long>133.5</long>
                <lat>-12.0</lat>
                <long>134.0</long>
                <lat>-12.0</lat>
            </dsgpolyo>
        </place>
        <place>
            <keyword thesaurus="auslig_topo250k_names"> Wessel Islands
            </keyword>
        </place>
    </spdom>
</descript>
```

Figure 3. Example of Metadata written in SGML using the ANZLIC DTD

Metadata Systems in ICZM

Several IT solutions and web catalogues are now available for improving data access using metadata, especially within general interest domains (e.g. environmental information) or for specialised data types (e.g. geodata or remotely-sensed images). There are however very few validated examples of metadata catalogues specifically suitable for ICZM. Furthermore, many such systems have failed to appraise the added value of providing metadata systems for data consumers and producers, the long-term economic viability of metadata systems and the sustainability and benefits for the coastal zone community that an effective metadata system could provide.

There are five key factors that limit the development of an effective ICZM metadata infrastructure[1]:

1. There is an insufficient consideration of the effectiveness of these systems to support real users in their established ways of working. Many technologically sophisticated solutions for data accessibility and exchange interfere with proven professional practices and introduce excessive requirements to the users.
2. Making data accessible introduces overhead costs to data holders (such as compilation, translation and maintenance of metadata records), often without clear financial or other benefits. This is a crucial issue for small-medium organisations, especially for those who do not have an institutional mandate for data sharing. Minimising overheads and identifying business models that maximise potential benefits for data holders is a fundamental incentive to improving data sharing practices.
3. In the past there has been an insufficient attention to the costs of ownership of systems that manage data flows and permit data sharing and accessibility. Issues such as the cost-benefit ratio of data and information services, economic viability and long-term economic survival of the systems have been often neglected.
4. Even when all these elements are in place, it is still difficult to identify the type and extent of positive impacts on ICZM. The ability to identify what sectors and activities benefit most (e.g., public v private sector) is fundamental to designing systems that target the right users and uses.
5. Finally, the trustworthiness of these systems largely relies on the quality of the data that it contains (e.g., matching between data and metadata). The lack of auditing schemes and quality protocols that ensure this quality may shrink the user pools and, in the long term, undermine the survival of the system.

These factors need to be considered in any system that acts to provide information to individuals and organisations in the coastal zone.

THETIS

THETIS[2] is an EC research project that plans to develop a realisable system that embraces the concept of information provision presented earlier in Figure 1. The

---

[1] This is of course assuming an agreement can be obtained from the data owner to make the data available through and information system. The legal regime and protection of intellectual property associated with electronic data transfer is still being defined and data owners are naturally wary of providing their data.

THETIS system seeks to address the frequent requirement of scientists, engineers and decision-makers to access, process and subsequently visualise data held at different locations. The objective of the THETIS project is to build an advanced integration system for transparent access and visualisation of data repositories, via the Internet using the World Wide Web (WWW). The system is planned to have the following features:

- On-line data access using WWW technology.
- Straightforward plug-in / publishing capability both for models and data sets.
- Efficient search for data and models via appropriate metadata descriptions and a distributed search engine.
- Flexible data integration using mediation techniques.
- Dynamic invocation of models to produce data on demand.
- Interactive data visualisation using GIS technology.
- Access level control.

Structure of the system
The architecture of the THETIS system connects various users via the Internet to a distributed collection of information systems. The system has a number of components:

- Data Repositories
- Search Engine
- Metadata
- Retrieval Engine
- Wrappers
- GIS

The data sets and models are stored in the data repositories. Each has an associated metadata form. The search engine uses the metadata to locate the data sets and models and the results of a search are returned to the user via the interface. To get a data set or to invoke a program, the location information is passed from the search engine to the retrieval engine. This retrieves the data or invokes the model, via the use of wrappers. The results are returned to the user via the GIS interface. The THETIS architecture is shown in Figure 4[3] and an example of the system operation is given in Figure 5.

THETIS Metadata
The metadata used by the THETIS project relies on the Federal Geographic Data Committee (FGDC) standard, which has been both extended and refined to suit the requirements of this project. Existing metadata descriptions can also be translated for use in the THETIS system. The metadata index is updated incrementally and does not involve human intervention. Search requests can be submitted to any server of the THETIS system from any location connected to the Internet.

---

[2] www.hrwallingford.co.uk/projects/THETIS
[3] In the diagram 'SQL' = 'Structured Query Language', a standard method for querying databases.

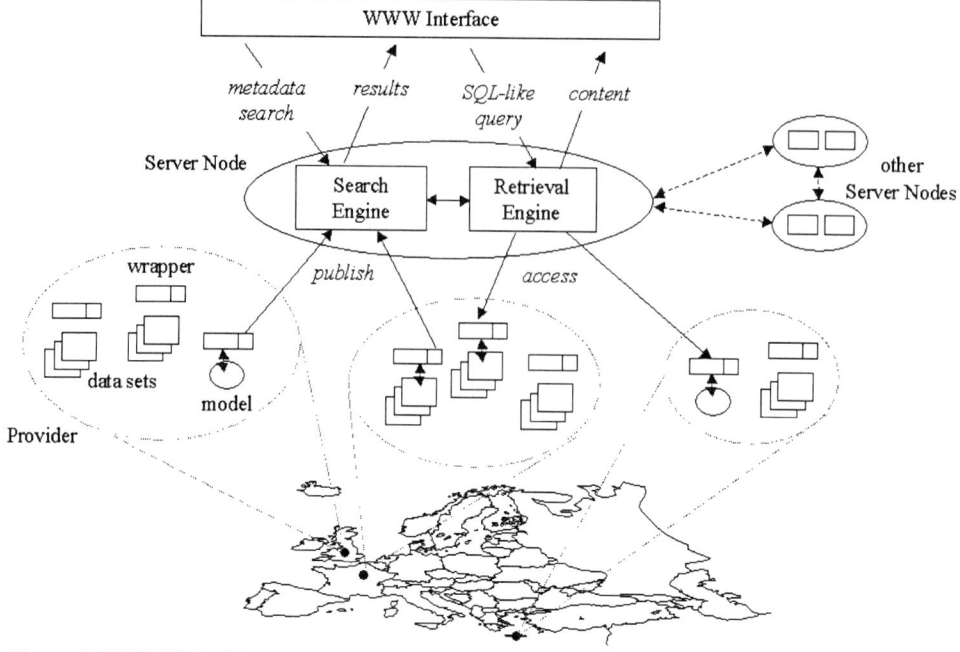

Figure 4. THETIS Architecture

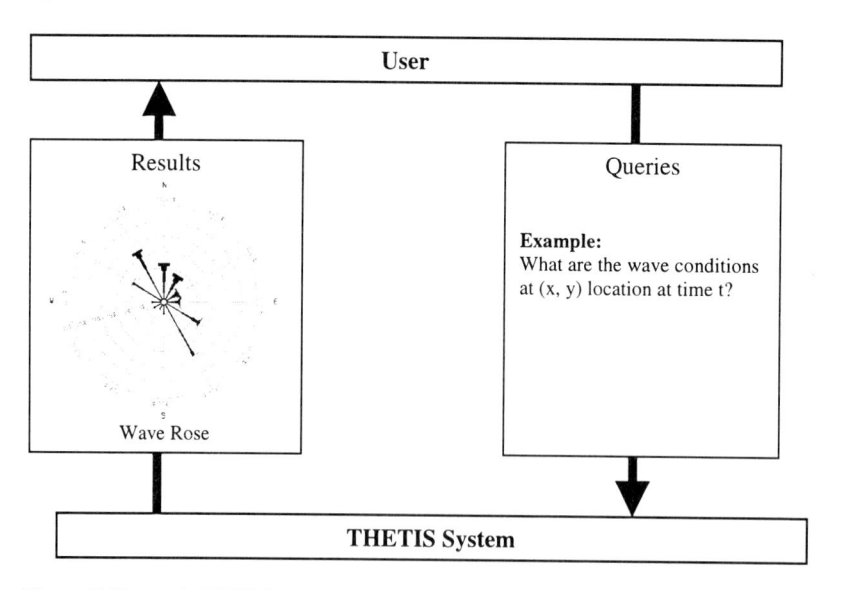

Figure 5. Example THETIS Operation in the Sea Defence Scenario

The concept of 'Wrappers'

Data and models are plugged into the system using special software adapters called 'wrappers' that export a relational view of data, independently of the actual underlying storage technology (e.g. files and databases). Model wrappers export an interface through which simulation programs can be invoked remotely. Wrappers allow data to be viewed interactively by combining data sources. They also allow data to be generated on demand by combining data sources and models. Both the data views and computed data can be published into the system just like ordinary data in order to be made available to others. To publish, wrappers and metadata are required for the newly created data set.

Scenarios to illustrate the system

Three coastal management scenarios are going to be used to demonstrate the functionality of the THETIS system. These scenarios combine the available data and models to produce new information that is displayed to the user. The scenarios can be summarised as follows:

Table 2. THETIS Demonstration Scenarios

Scenario	Features
Waste Transport	<ul><li>Simulation of pollution point sources.</li><li>Computation of general circulation data.</li><li>Computation of the concentration of effluents based on general circulation data.</li><li>Visualisation of the transport of pollutants.</li></ul>
Tracking of Sea Structures	<ul><li>For the study of the dynamic of oceans.</li><li>Extraction of oceanographic structures from satellite images through stepwise image processing.</li></ul>
Nearshore and Offshore Waves	<ul><li>For coastal defences, coastal and offshore structures, flood prevention, etc.</li><li>Calculation of the wave climate based on historical wind and wave data.</li><li>Estimation of directional (2D) spectra of surface waves based on heave-pitch-roll time series.</li></ul>

Within the three scenario areas, THETIS provides a system that can integrate data from different sources with processing models. The sea defence scenario is designed to demonstrate how a pan-European data set, in this case a wave data archive, can be queried for any spatial or temporal location. Dependent on the user-request, THETIS applies background processing to extract the relevant data and then invokes the user-selected visualisation routine.

CONCLUSIONS

It can be considered that the 'Holy Grail' of an information provision for ICZM is to enable a user to query a system with 'tell me about x' and data on 'x' is returned at a level appropriate to the user. Providing this kind of service is technically possible, but there is still a way to go before this type of technology is common-place, particularly to ensure that the system is capable of being populated with all the necessary data sets

Within the next few months HR Wallingford will be testing an internet-based service whereby a user can make an query of the form 'what is the wave climate at x,y over period t' and the user is returned with a wave rose of the wave climate, with an option to download the data. This approach is bringing the concept of the 'data shop' closer to the customer, albeit in one tightly define application area. This, however, can be regarded as the first steps towards the system 'ideal'

Technology can provide a sophisticated level of processing that facilitates advanced searching, retrieval and processing of data. This can help overcome barriers to data use by minimising format incompatibilities, data delivery times and improving the ability to determine if data is going to be useful.

However, the amount of human effort required so that data can be accessed in such a fashion should not be understated. If we consider the THETIS system, human operators are still required to provide authorship and publishing of both wrappers and metadata. Writing wrappers is more a less a well defined task, however writing metadata is open to human 'creativity'. The adoption of metadata standards and metadata editors (these generate SGML in a similar way that a web-page editor generates HTML) make the task of producing metadata less error-prone, but content is still largely up to human interpretation.

Metadata management and metadata interoperability are crucial to improve data diffusion. Without adequate metadata, a system such as THETIS would not be able to operate, negating the sophistication of the processing technologies. This is an important consideration as UK experience indicates that for many organisations thorough metadata generation is not a priority. This can partly be attributed to the lack of guidelines available advising of approaches to use and promoting a culture of metadata generation.

In conclusion, the ability to generate more useful information more quickly through improved access to data sets is important for improved management in the coastal zone. To advance the present situation and take advantage of technological developments, there needs to be increased emphasis in the coastal management community on both using and producing metadata. National guidance on this topic would benefit this process.

REFERENCES
CEO (Centre for Earth Observation), 1996b, Survey of current practices and standards for metadata management; Volume A: Survey and analysis of practices in 20 EU organisations, Document Reference: CEO/US/1400/217, Issue 1.1.
CIRIA, 1999, Maximising the use and exchange of coastal data: A guide to best practice (in press).
EC (European Commission), 1999, Lessons learnt from the European Commission's Demonstration Programme On Integrated Coastal Zone Management, ISBN 92 828 6471 5 DGXI
OECD (Organisation for Economic Co-operation & Development), 1993, Coastal Zone Management – Integrated Policies.

Maximising the use and exchange of coastal data: A guide to best practice

P SAYERS*, K MILLARD*, D LEGGETT**
*HR Wallingford Ltd
** CIRIA

ABSTRACT
There are over 250 Government Agencies and local coastal authorities in the UK with responsibilities in the coastal zone; most of whom gather data. In addition, universities, research institutes, port and harbour authorities, consulting engineers and environmental scientists collect coastal data. The reasons behind collection are disparate and different data standards and formats are used. The result is that many thousands of different datasets are collected, however, many are incompatible.

Existing institutional and organisational policies on pricing, copyright and confidentiality also inhibit data sharing. Data sharing is further restricted by issues of data quality and technological aspects such as archive systems, the content of multi-media datasets (such as paper, micro-fiche, electronic) and cataloguing methods.

To address these issues, and maximise data use and value by improved data exchange, CIRIA commissioned HR Wallingford to undertake an investigation into methods to encourage common data brokerage and to facilitate data sharing within and between sectoral groups and organisations.

This paper presents the findings of this research. In particular, it explores methods for improving existing exchange of data and removing the barriers that currently prevent data exchange. Issues relating to both policy and technology are explored to maximise both primary and secondary use, data exchange across activity sectors and of future reuse.

INTRODUCTION
Many organisations operating in the coastal zone have identified that they have insufficient data upon which to base management decisions. This is partly because the data simply do not exist but is more often because existing data are difficult to find, of insufficient quality, in the wrong format or simply too expensive. It is recognised, however, that common data needs exist between organisations and that improved exchange would reduce the need for new collection and reduce duplication of effort.

The inefficient way in which most data are collected and shared today is a legacy of collection and provision routes that have evolved in organisations and industry sectors over many years. These routes encompass differing data collection methods, data processing and storage technologies and data control policies. The increased adoption of integrated management initiatives on the coast, such as Shoreline Management Plans and Estuary

Management Plans, have highlighted the restrictions that current data management practices place on data sharing.

If all organisations operating in the coastal zone were to adopt policies to maximise the use and exchange of coastal data operating costs would reduce and efficiency would increase. It would also facilitate better decision making in all aspects of coastal management.

Purpose and scope of the best practice guide
In tandem with the Best Practice Guide (CIRIA, 1999b) this paper sets out to support individuals and organisations who work in coastal areas by aiding them with the management of their data. It seeks to summarise how data suppliers can make their data more accessible to other organisations and how data customers can best make use of existing datasets.

In summary, the aims of the paper are to:

- Produce guidance and recommendations for organisations who undertake coastal monitoring and coastal data collection on consistency of data and mechanisms for data exchange between organisations.
- Encourage the concept of the common brokerage of coastal data and to facilitate data sharing within sectoral groups and between organisations.
- Improve understanding of the mechanisms and issues that affect data use and exchange.
- Improve awareness of data types and availability.

This paper summaries the key policies and technologies report in the best practice Guide that may be employed to maximise the use and exchange of coastal data to support these aims and the interested reader is referred to the Guide itself (CIRIA, 1999b) for more details.

Data use and exchange
Data sharing leads to data integration and hence to more meaningful analysis and better information. It also results in reduced duplication of effort and, hence, cost savings. If achievable, the integration of diverse data provision activities, serving different functions in the coastal zone, will, in effect, create a comprehensive data set for the strategic management of the coast. Unlike the USA, there are no centralised initiatives at UK government level to support the use and exchange of coastal data; it is therefore important that the initiative is taken by individual organisations operating in the coastal zone.

The main benefits of maximising data exchange and re-use are:

- Reduced overall operating costs.
- Improved relationships with customers and suppliers.
- Increased revenue through sales of data products.
- Improved analysis and understanding of the environment.
- Aid to sustainable development.

Naturally, these benefits need to be justified against the financial cost to any single organisation of making more of their data available and any loss of competitive advantage.

Data exchange and re-use is bounded by a combination of *policy* and *technological* issues. Examples include the choice of storage medium, the format of the data, archive policies and pricing agreements. These issues are inter-related; for example, advances in digital media have led to changes in copyright agreements. There are many other key issues that have a bearing on the use and exchange of data and these are shown graphically in Figure 1.

The effect these issues have on data exchange vary between the individual data users and suppliers. They also change over time. What may be an 'annoyance' today may completely prevent data exchange in 20 year's time. It is important to remember that 20 years ago the desk top computer was a rarity and 5¼" disks were the latest exchange medium.

It is interesting to note that in recent years many reports and publications have concluded that effective sharing of data is fundamental to the development of the integrated management techniques such as Coastal Zone Management, Estuary Management and Shoreline Management (OECD 1993, English Nature 1993, DoE 1995b). However, whilst these reports have stressed that maximising data exchange is something that *should* be done, they do not prescribe *how* it should be done. This paper sets out to present best practice on how maximising data exchange.

Each sector operating in the coastal zone requires information to underpin the decision-making of their constituent organisations. This information is obtained by processing data. Despite disparate information needs, however, sectors often have similar data requirements to satisfy their information requirements. For example, both Tourism and Coastal Defence require data on hydrodynamics, terrestrial biology and socio-economics, amongst others.

At the project workshop (CIRIA, 1999a) it was established that the most significant factors that affect data use and exchange are:

- Data customers are unable to find data.
- Data customers are unable to assess the quality of data.
- Data suppliers are unsure what customers require.

It is interesting to note that factors such as 'incompatible data formats' were not considered significant barriers to data exchange, providing the format was known. For example, if the data were held on Exebyte tape and the potential data customer did not own an Exebyte tape reader, then providing the potential customer could assess the value of the tape's content, a way around the barrier would be found. One approach would be to ask an organisation with access to an Exebyte tape reader to extract the data.

110 COASTAL MANAGEMENT

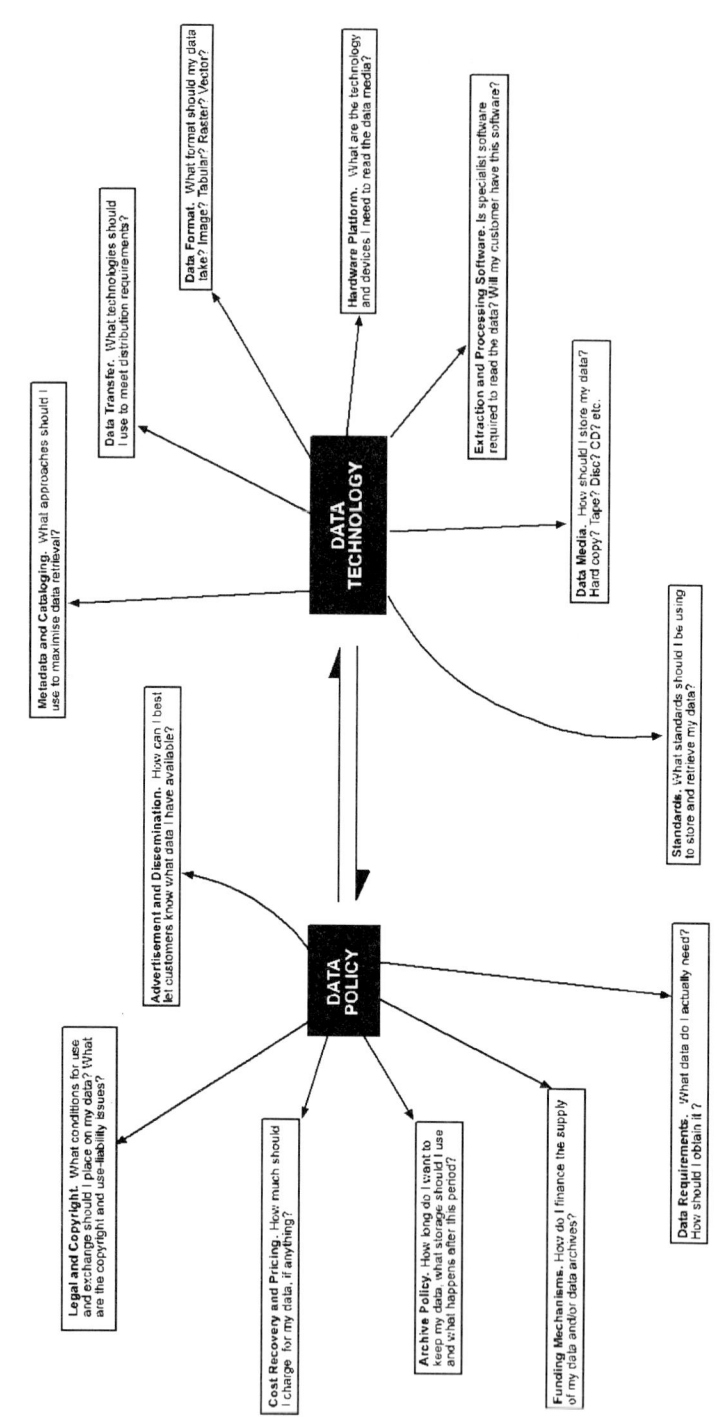

Figure 1   Factors affecting data exchange

Information for Coastal Management

Figure 2 illustrates the key coastal zone sectoral areas and the UK institutions active within these sectors. Together, these organisations are the suppliers and customers of coastal data in the UK.

Each of these sectors require information to underpin decision making. This information is obtained from processing data. Despite disparate *information* needs however, sectors often have similar *data* requirements. For example, both the Tourism and Coast Defence sectors require data on hydrodynamics, terrestrial biology and socio-economics amongst others. Similar common data requirements can be identified between other sectors. This is demonstrated in Figure 3 that shows the typical data required in each of the main sectors and how the data requirements overlap for the Tourism and Coastal Defence sectors.

Breaking down the barriers to data exchange

Both data customer and data suppliers need to ensure that data policy and data technology issues operate to advance data exchange rather than manifesting themselves as barriers. This is not an impossible task, and one that can largely be accomplished through:

- Actual rather than perceived data and data processing needs.
- Integration of data exchange practice into existing organisation practice.

This means that if it is essential that data customers carefully appraise the data requirements (to satisfy their own and their clients' information needs) and communicate these needs accurately to data suppliers. For data suppliers, this means fully understanding what your customers' data needs really are. However, unless the business processes of an organisation are effective, any changes in data re-use and supply will have limited effect on improved business operation or wider data exchange.

112 COASTAL MANAGEMENT

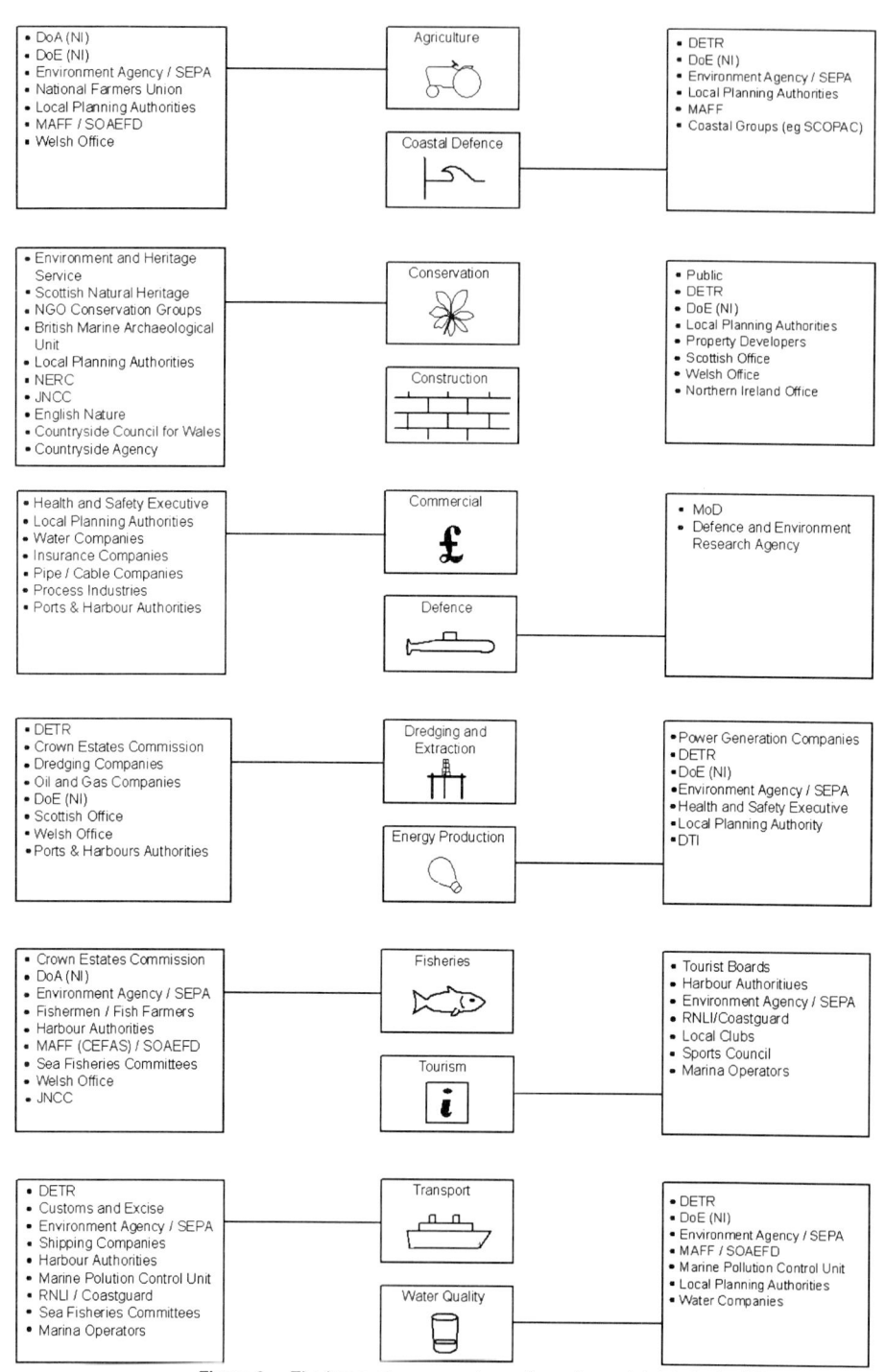

Suppliers	Sector	Customers
• DoA (NI) • DoE (NI) • Environment Agency / SEPA • National Farmers Union • Local Planning Authorities • MAFF / SOAEFD • Welsh Office	Agriculture	• DETR • DoE (NI) • Environment Agency / SEPA • Local Planning Authorities • MAFF • Coastal Groups (eg SCOPAC)
	Coastal Defence	
• Environment and Heritage Service • Scottish Natural Heritage • NGO Conservation Groups • British Marine Archaeological Unit • Local Planning Authorities • NERC • JNCC • English Nature • Countryside Council for Wales • Countryside Agency	Conservation	• Public • DETR • DoE (NI) • Local Planning Authorities • Property Developers • Scottish Office • Welsh Office • Northern Ireland Office
	Construction	
• Health and Safety Executive • Local Planning Authorities • Water Companies • Insurance Companies • Pipe / Cable Companies • Process Industries • Ports & Harbour Authorities	Commercial	• MoD • Defence and Environment Research Agency
	Defence	
• DETR • Crown Estates Commission • Dredging Companies • Oil and Gas Companies • DoE (NI) • Scottish Office • Welsh Office • Ports & Harbours Authorities	Dredging and Extraction	• Power Generation Companies • DETR • DoE (NI) • Environment Agency / SEPA • Health and Safety Executive • Local Planning Authority • DTI
	Energy Production	
• Crown Estates Commission • DoA (NI) • Environment Agency / SEPA • Fishermen / Fish Farmers • Harbour Authorities • MAFF (CEFAS) / SOAEFD • Sea Fisheries Committees • Welsh Office • JNCC	Fisheries	• Tourist Boards • Harbour Authoritiues • Environment Agency / SEPA • RNLI/Coastguard • Local Clubs • Sports Council • Marina Operators
	Tourism	
• DETR • Customs and Excise • Environment Agency / SEPA • Shipping Companies • Harbour Authorities • Marine Pollution Control Unit • RNLI / Coastguard • Sea Fisheries Committees • Marina Operators	Transport	• DETR • DoE (NI) • Environment Agency / SEPA • MAFF / SOAEFD • Marine Pollution Control Unit • Local Planning Authorities • Water Companies
	Water Quality	

Figure 2   The key customers and suppliers of coastal data in the UK

Coastal atmosphere	Agriculture	Soil
• Wind parameters • Heat flux • Moisture flux • Air temperature • Relative humidity • Atmospheric pressure • Precipitation rate • Cloud cover		• Soil moisture • Nutrient levels • Contaminant levels • Water content

Coastal Defence

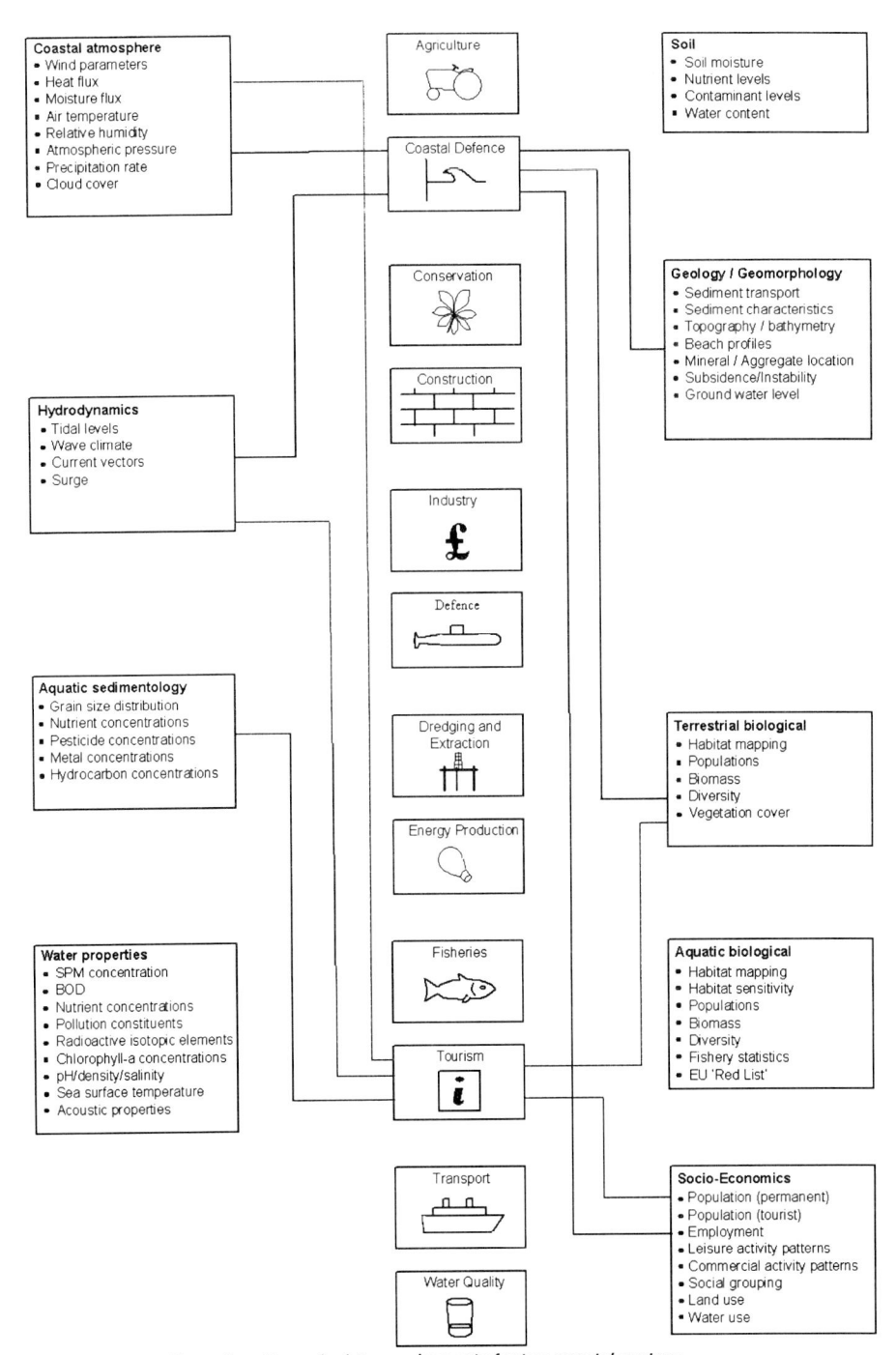

	Conservation	Geology / Geomorphology
		• Sediment transport • Sediment characteristics • Topography / bathymetry • Beach profiles • Mineral / Aggregate location • Subsidence/Instability • Ground water level

Construction

Hydrodynamics		
• Tidal levels • Wave climate • Current vectors • Surge		

Industry

Defence

Aquatic sedimentology	Dredging and Extraction	Terrestrial biological
• Grain size distribution • Nutrient concentrations • Pesticide concentrations • Metal concentrations • Hydrocarbon concentrations		• Habitat mapping • Populations • Biomass • Diversity • Vegetation cover

Energy Production

Water properties	Fisheries	Aquatic biological
• SPM concentration • BOD • Nutrient concentrations • Pollution constituents • Radioactive isotopic elements • Chlorophyll-a concentrations • pH/density/salinity • Sea surface temperature • Acoustic properties	Tourism	• Habitat mapping • Habitat sensitivity • Populations • Biomass • Diversity • Fishery statistics • EU 'Red List'

	Transport	Socio-Economics
	Water Quality	• Population (permanent) • Population (tourist) • Employment • Leisure activity patterns • Commercial activity patterns • Social grouping • Land use • Water use

Figure 3   *Example data requirements for two coastal sectors*

114 COASTAL MANAGEMENT

So, how is maximum use and exchange of data achieved?

In order to achieve maximum use and exchange of data five principles of good coastal data management have been developed. These are based on principles adopted by the British Standards Institute for the management of electronic documents (Mayon-White *et al.* 1997) (see Figure 4). Recognising that different sectors and organisations involved with coastal data will have existing procedures for data management already in place, the principles are technology independent. This ensures they will remain valid in the future.

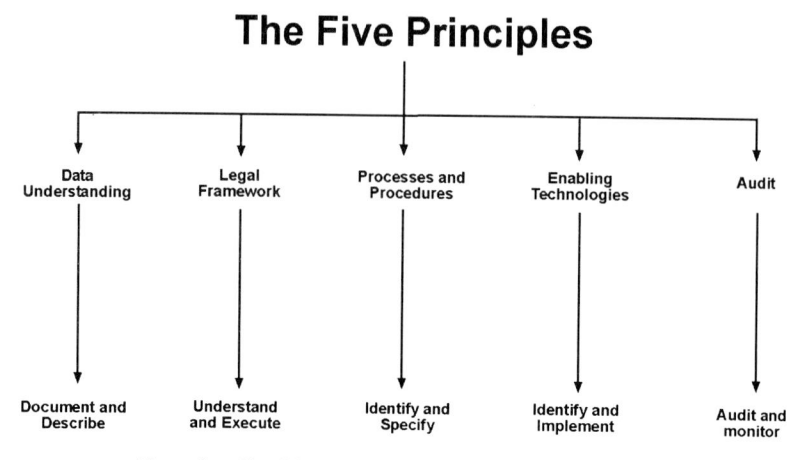

The Five Principles

```
                    The Five Principles
                            |
   ┌────────────┬───────────┼───────────┬────────────┐
   ▼            ▼           ▼           ▼            ▼
  Data        Legal      Processes   Enabling      Audit
Understanding Framework  and         Technologies
                        Procedures
   │            │           │           │            │
   ▼            ▼           ▼           ▼            ▼
Document and Understand  Identify and Identify and Audit and
 Describe    and Execute   Specify     Implement    monitor
```

Figure 4 *Five Principles of good Data Management*

The issues and practices surrounding the 'Five Principles' shown in Figure 4 are summarized below:

Principle 1: Data understanding

Fundamentally and primarily, any organisation must appraise its data requirements to ensure they match their, or their customer's information needs. It is not sufficient to assume that because certain data are collected or generated they will always be needed. Data demands need to be expressed explicitly, not only to organisations who may be subcontracted to provide the data but also to other organisations operating in the coastal zone. These organisations may be capable of contributing towards a joint data collection exercise.

Data must always be provided with a description of their content that must be identifiable to the data. This description is referred to as metadata and is singularly one of the most important requirements to both streamline data management and improve data exchange.

Finally, organisations should take steps to establish methods to disseminate widely data catalogues in their possession or identify organisations that could do this on their behalf.

Principle 1: Data understanding

> **Recognise, understand and describe all data used, needed and available**
>
> - Appraise data requirements to satisfy information needs.
> - Recognise all organisations as potential data suppliers or data customers.
> - Communicate data requirements and data availability with other organisations.
> - Ensure all data are accompanied at all times with appropriate metadata.

Principle 2: Legal framework

Legal issues such as copyright and licences are essential to the economy of data provision, however different data suppliers stipulate different agreements. This is confusing to the data user and data suppliers should aim to make agreements for the use of their data as straightforward as possible. Confidentiality of data also needs to be respected, and organisations supplied with data need to inform all relevant staff of limitations on data use. Finally, all staff involved with the collection and processing of data need to be made aware that they have a 'duty of care' to preserve data. Everybody is responsible to ensure that the lifecycles of any data are managed effectively.

Principle 2: Legal framework

Understand legal issues and execute responsibilities
• Ensure conditions for the use and exchange of data are clearly defined between customer and supplier
• Ensure all staff in an organisation operate a "duty of care" towards the management of data.

Principle 3: Processes and procedures

The procedures and processes within an organisation that are part of a data processing chain should be described and documented. This means that if an organisation is a collector of data, it needs to be stated what techniques are used. Similarly, if an organisation generates data using modelling or data fusion methods, such procedures also need to be documented. In practice, this documentation is usually incorporated in reports or notebooks but steps should be taken to ensure that as much interaction as possible is incorporated in metadata.

These procedures must also make the data life cycle fully visible. An organisation must be able to account for all the data in their possession and have a policy on ownership. In particular, an organisation who provides data always needs to establish with its client the fate of all data used as part of a project and not just those requested as a final deliverable.

Principle 3: Processes and procedures

Identify and specify organisation processes and procedures
• Document and describe all organisation procedures associated with the data processing chain relevant to your organisation.
• Identify, specify and implement organisational procedures that incorporate management of data lifecycles.

Principle 4: Enabling technologies

Organisations must apply appropriate technologies for the storage, transmission and management of data. This does not mean the organisation has to constantly keep up with the latest technological developments, but it should recognise that technology does become obsolete and thus make informed decisions about upgrade programmes.

Working with data suppliers and data customers can help minimise the costs associated with technology management as well as improving data use and exchange. For example, agreements on data formats and software platforms reduces risk of technological obsolecence and minimises data conversion time.

Finally, as data are increasingly stored in electronic format, organisations must ensure that confidential data are not illegally duplicated or accessed, and that the data are effectively protected on its media and not allowed to deteriorate to prevent its future use.

Principle 4: Enabling technologies

Identify and implement appropriate technologies for data
• Identify technologies that are compatible with data supplies and data customers. • Monitor changes in data transfer and storage technologies and apply those appropriate to business processes. • Appraise the level of security and protection attributed to the storage of data

Principle 5: Audit

Audit is an essential part of any practices for maximising the use and exchange of data. It enables the benefits of the process to be quantified and areas of improvement identified. If the organisation already has a formal audit procedure, then procedures to improve data exchange should be incorporated within it. If the organisation has no procedures in place then the organisation should establish an audit process.

Principle 5: Audit

Audit and monitor process for data use and exchange.
• Identify indicators to monitor the success of data exchange • Review and monitor.

Conclusions

Monitoring of the coastal zone continues to increase. However, the full potential data in the decision making process is, at present, not realized. This paper has set out the five key principles that will aid organisations operating in the various activity sectors to share data in an efficient and reliable fashion.

The increased use of data, that improved exchange brings, will directly support better decision making in all aspects of coastal management and commercial activities. To be effective, however, it is also clear that a desire to maximise the use and exchange of data can not be viewed as being separate to these operations.

REFERENCES AND BIBLIOGRAPHY

CIRIA, 1996, *Beach Management Manual*, CIRIA Report 153.

CIRIA, 1999a, Proceedings of workshop into maximising the use and exchange of coastal data held as part of the MUSEC project at HR Wallingford, 19 February 1999.

CIRIA, 1999b, Maximising the use and exchange of coastal data: A guide to best practice (in press)

DoE (Department of the Environment) 1995, *Coastal Zone Management – Towards Best Practice*, Prepared by Nicholas Pearson Associates, October 1996

English Nature, 1993, *Estuary Management Plans – A co-ordinators guide* published by English Nature, ISBN 1 85716 121 1

Mayon-White, W and Dyer B, 1997 *Principles of Good Practice for Information Management*, Version 2.0 IMDA, London School of Economics, Published by British Standards Institute, ISBN 0 580 26855 1

OECD (Organisation for Economic Co-operation & Development), 1993, Coastal Zone Management – Integrated Policies.

The Development of a Framework for Estuary Shoreline Management Plans

DR N I PONTEE and I H TOWNEND
ABP Research & Consultancy Ltd, Southampton, England

ABSTRACT
This paper describes an approach to estuary shoreline management planning (eSMP) based upon the shoreline management planning (SMP) approach which has been used extensively around the UK coast.

The proposed method comprises two parallel strands, one following a functional approach and the other focussing on resources. Both approaches involve using a cause-consequence model to identify the temporal and spatial scale of any changes in the estuary system that may result. The estuary SMP process may be summarised as follows:
(i) Identify functional requirements to develop "design" concept.
 Evaluate the size, shape and location of the changes that are likely to result.
 Assess the implications for **estuary resources.**
(ii) Use resource interests to define an initial set of SCDO's.
 Evaluate the size, shape and location of the changes that are likely to result.
 Assess the implications for the **estuary system.**
(iii) Compare the options against the management objectives and consult to identify preferred SCDO's.
(iv) Map SCDO's as the management units for the estuary.

The proposed methodology provides a consistent procedure for defining SCDO's, that can be revisited as the science base advances, new data becomes available, or new infrastructure needs are identified. It therefore provides a sound framework on which to develop a long-term interactive management plan.

INTRODUCTION
This paper describes an attempt to adapt the coastal SMP process for use within an estuary. The resulting approach is referred to as an estuary shoreline management plan (eSMP). The aim of the eSMP is to provide a consistent procedure for defining SCDO's in estuaries which can form the basis of a management plan over the medium to long-term. The first section of the paper describes the coastal SMP procedure. The next section describes the differences between estuaries and open coasts and the requirements of an eSMP. The recommended approach to eSMP development is then described in the final section.

THE SMP PROCESS
MAFF (1995) divides the production of coastal SMPs into two stages:

1. Data collation, analysis and setting overall objectives.
2. Plan preparation.

The aim of Stage 1 is to gather and analyse information and identify all those with an interest in the area. This enables management units to be defined, and management objectives to be set for the shoreline. Stage 2 involves the definition of management units and appraisal strategic coastal defence options. The available Strategic Coastal Defence Options (SCDO's) for each management unit are:
(i)   do nothing;
(ii)   hold the existing defence line by maintaining or changing the standard of protection;
(iii)   advance the existing defence line;
(iv)   retreat the existing defence line.

Any SCDO studied must be sustainable and compatible with the preferred options identified for adjacent management units, as well as the processes at work within the sediment cell. Each option must also be evaluated in economic and engineering terms.

Within stage 2 the SMP process the definition of management units is of fundamental importance. MAFF (1995) defined a management unit as a length of shoreline with coherent characteristics, in terms both of natural coastal processes and land use. A number of different methods have been used to define management units and set SCDO's (see Townend *et al* 1996; Halcrow, 1991). However, a common problem in existing methods SCDO's is that management units tend to be controlled by resource issues on the developed coast, whilst on the undeveloped coast units are more likely to reflect the natural process and attendant geomorphological features. Thus the status quo tends to prevail.

A fuller description of the SMP process can be found in Purnell (1996) and a consideration of some of the technical issues involved in the preparation of SMP's is given by Townend *et al.* (1996), and Leafe *et al.* (1998).

REQUIREMENTS FOR AN ESTUARY SMP
Estuaries have a number of characteristics which distinguish them from open coasts and which are especially relevant to the adaptation of the SMP process (HR *et al.,* 1996). These include:
- The more complex water movement due to the enclosed channel morphology and the influence of freshwater and saline inputs.
- The greater degree of complexity in sediment transport pathways, which reflects both the complexity of water movements and the fact that sediment can be supplied from both marine and freshwater sources.
- The high degree of sediment reworking and the juxtaposition of erosional and depositional shores.
- The finer grades of sediment in transport, typically mud and sand, as compared to sand and gravel on coasts.

Within an estuary, form and process are inextricably linked, there are no obvious dependent and independent variables or clear cause-effect hierarchics. For example, although the size and shape of an estuarine channel is a response to tidal processes, the tidal discharge is itself dependent on the morphology of the estuarine channel since this determines the overall tidal prism. This interaction results in the potential for small changes to have far reaching effects.

Whilst "process units" have been identified within a sub-cell on the open coast, for an eSMP, it is more appropriate to take the whole estuary as the basic "process unit". This demands a wider-scale, longer-term approach, driven by an understanding of physical processes. Developments within an estuary need to be considered in terms of both their local and estuary wide impacts.

The approach that needs to be adopted requires the investigation of the interactions between process and form, which give rise to the functional behaviour of the estuary system. Interventions, such as coastal defences, have the potential to alter estuarine form and/or processes, and in this way affect all parts of the estuary to some degree. Of critical importance is the need for the management framework to avoid focusing in on individual units, as tends to happen on the coast. Within the estuary there is a duality required, by which the manager considers not only local actions - within a management unit - which achieve a particular objective, but also whether there are actions that might be taken remotely that could equally achieve the objective. In all cases, it will be necessary to consider the impacts of a particular management action at all levels of the system. This leads to the concept of estuary "design" to meet the functional requirements of the users, whilst at the same time seeking to sustain the desired natural function of the system.

In an estuary the potential for small changes to have far reaching effects means there is a the need for a greater appreciation of the likely geomorphological impacts of any given defence strategy. Consequently, it is unlikely that the method of defining management units on the basis of resource and process units will be successful, due to the complexity of interaction within the estuary system. These interactions apply to both the needs of the estuary users as well as the physical processes operating. Thus in estuaries it is suggested that management units are best defined on the basis of unique SCDO's.

The activities of users can interact directly with the estuary processes and indirectly with the flood defences or coast protection. To achieve sustainable use of the estuary, a plan for coastal defence needs to consider effects of navigation, nature conservation, urban areas, etc for the whole estuary. This requires an understanding of (i) how the estuary functions as a whole, (ii) the various user objectives within the estuary, (iii) the likely consequences of change, both natural and man-made.

PROPOSED METHODOLOGY
As for coastal SMP's the whole process of the production of an eSMP can be split into two stages, each composed of a number of steps. The first stage of the procedure for an estuary, is virtually unchanged from coastal SMP's. Stage 1 involves data collation, analysis and the setting overall objectives. It involves three main steps:
(i)   identify all those with an interest in the area;
(ii)   collate and analyse existing data on all the key issues;
(iii)   set management objectives for the Plan area which form the basis for the appraisal and development of strategic coastal defence options.

Within an estuary, it is envisaged that there will be a set of broad objectives for the estuary as a whole. Some typical objectives drawn from MAFF guidance and previous SMP's are summarised in Table 1. There may then be some localised objectives, not linked to management units (because they will not be defined *a priori*), but defined over lengths of shore for which there is some specific functional requirement.

Table 1: Possible Management Objectives

Type of objective	Broad
Scope	For the estuary as a whole.
Examples	To adopt an estuary shoreline management plan (eSMP) consistent with the dominant processes and environmental constraints.
	To explore the opportunities to establish an estuary wide functional morphology, incorporating necessary human interventions and modifications.
	To ensure that the eSMP is based on sound economic and technical principles, which represent value for money.
	To identify opportunities for enhancement of landscape, amenity, conservation and the local economy.
	To inform the statutory planning process.
	To establish a programme and procedure for reviewing eSMP.
	To liase with and develop close working relationships with relevant authorities and organisations as necessary to work towards an integrated estuary management approach.
	To promote a greater public awareness and understanding of the estuary and how it influences local issues.
Type of objective	Local
Scope	For lengths of shore which have some specific functional requirement, they therefore to highlight specific local needs
Examples	To protect such things as local habitats, historic sites and access routes – these can be viewed as preferences which may, if necessary, be over-ruled by the estuary wide objective
	To protect an urban conurbation – in this case the local objective may represent constraints on what can be achieved in terms of the estuary wide objectives.

Stage 2 involves the definition of management units and the choice of SCDO's. The process is broken down into two parallel approaches with a series of steps in each (Figure 1). The first approach starts with the concept of establishing an estuary wide design on the basis of a functional accommodation of the processes, fits suitable SCDO's into this framework, and then moderates these by a consideration of resource issues. Another way of expressing this approach would be to say "How will the estuary evolve and, in an ideal world, how can we accommodate this?". The second approach is resource based. It starts with a choice of SCDO's based solely on resource issues, and then moderates these by consideration of the physical processes operating in the estuary. Another way of expressing this would be to say, "In socio-economic terms, what would we like to do?".

It is apparent that the two approaches are effectively the converse of one another. Both take into account resource and process constraints, although the weightings given to each differ. Each approach produces a map of SCDO's around the estuary. The final step of the process is to try and reconcile the outcomes of the two approaches. The SCDO's require testing against the management objectives, taking due account of both process and resource implications. It is likely that this will make extensive use of consultation, in order that:
(i) an improved understanding of the issues can be disseminated to as wide a group as possible; and,
(ii) that this will, in turn, engender a better appreciation of the need for longer term planning, which is not unduly constrained by the status quo.

The aim of this final exercise should, of course, be to develop a broad based acceptance of the Plan.

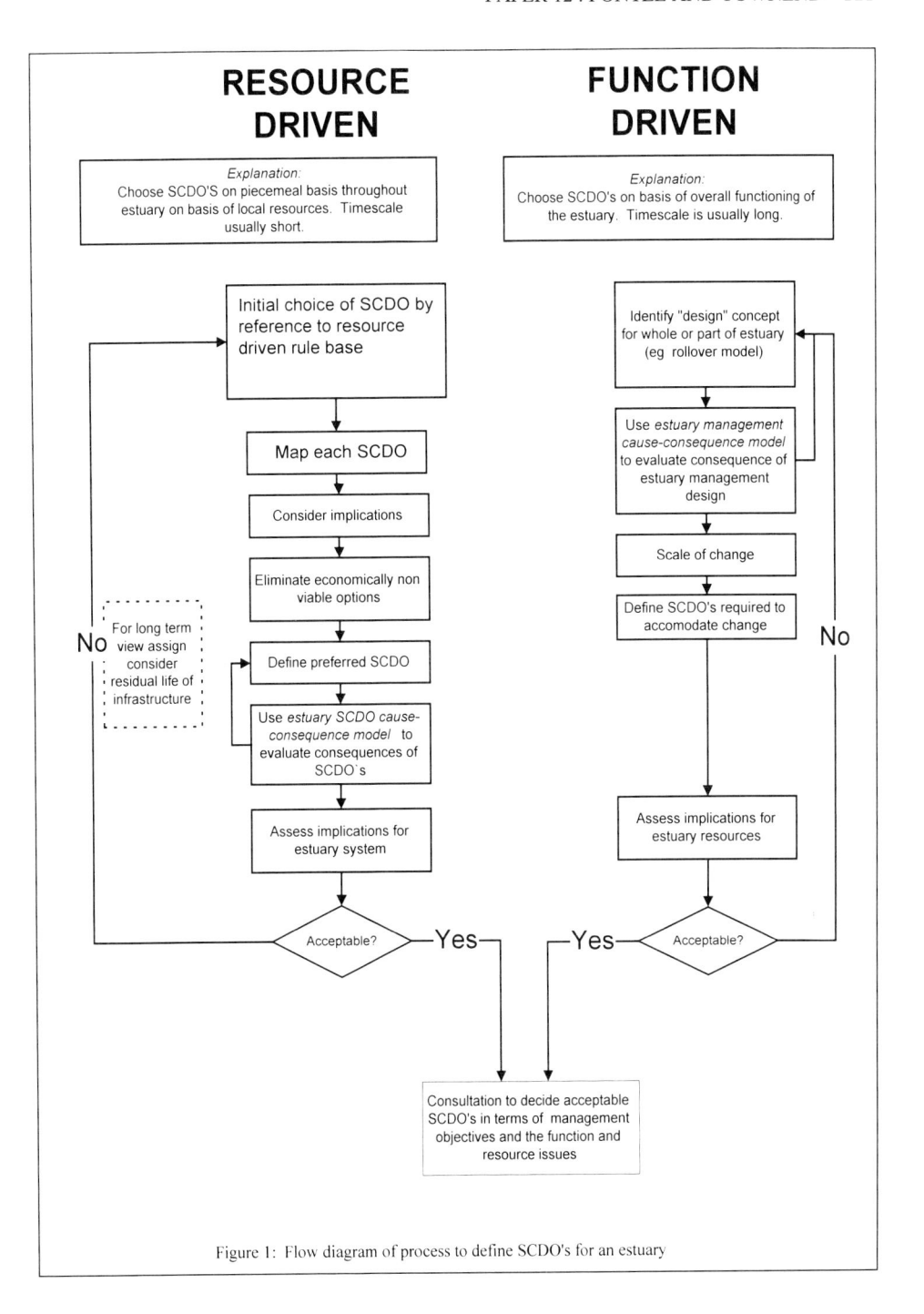

Figure 1: Flow diagram of process to define SCDO's for an estuary

(a) Function driven approach

This approach requires viable concepts to be identified and modelled to determine the spatial and temporal scales of change. This then provides a guide to where defences ought to be (or ought not to be!) and hence a set of SCDO's capable of delivering a particular estuary "design". The first step for the function driven approach, shown on the right side of Figure 1, is to refer to a conceptual design concept for the whole, or part of the estuary. One such design concept may be the rollover model for the landward transgression of the estuary proposed by Pethick (1996).

The next step is to evaluate how the system is likely to respond. This is explained conceptually through the *Estuary Management Cause-Consequence Model* (Figure 1) (Pontee and Townend, 1999). This model informs the user of the consequences of possible change in the estuary system, including energy, sediment and modifications to the estuary form through management actions. Each set of "causes" will have a spatial and a temporal scale as indicated in Figure 2. In combination these will invoke a response (or responses), the outcome of which may be at one or more spatial and temporal scales. An explanation of how the cause-consequence model and, in particular, the central response model is formulated, is given in a separate paper (Pontee and Townend, 1999).

It should be noted that under 'Management Actions' in Figure 2, sea defences represent a summary of the four SCDO's expressed in the *Estuary SCDO Cause-Consequence Model* (Figure 3). The model aims to inform the user of the scales of estuary change that will occur. Once the scale of change is known, SCDO's can be defined around the estuary, to accommodate these changes. It may well be that conflict arises here, since the function driven design concept may suggest SCDO's which are unacceptable on resource grounds. If this is the case, then it is necessary to return to the choice of SCDO's required to accommodate the design concept, or else reconsider the original design concept.

(b) Resource driven approach

For the resource driven approach, shown on the left side of Figure 1, the first step is to make an initial choice of SCDO's for the entire shoreline of the estuary, based upon a resource driven rule base (see Townend *et al.*, 1996). Once a range of SCDO's has been defined in this way, it is necessary to consider the implications of these SCDO's and eliminate those that are not economically viable. This then leaves a preferred SCDO, based purely on resource issues, for each length of the estuary shoreline.

In an estuary although process units are both areal and temporal, coastal defences are relatively fixed and linear. The areas at risk from flooding and/or erosion determine the location and extent of defence requirements. In the main, these relate to a particular length of shoreline and so, as with the open coast, can be described by linear units which run along each bank. However, simply defining the length over which a defence is required is not sufficient to enable interactions with the estuary system to be evaluated. To assess the likely outcome of any change we need to know both the transverse location and lateral extent of the defences. This is what an SCDO provides, at least in a spatial rather than structural sense. Thus it is necessary to assign SCDO's to lengths of the estuary. To begin with these SCDO's are defined by landward resource interests, and take little or no account of natural processes.

Such a resource based definition of SCDO's then needs to be tested within a functional model of the estuary system, to identify the spatial and temporal scales of any impact. Here again, the cause-consequence model, described above, serves this purpose. An iterative process may be required to identify a set of SCDO's which result in acceptable local and system wide impacts.

In order to take account of process issues, which may mean that some of these SCDO's are unacceptable, it is necessary to use the *Estuary SCDO Cause-Consequence model* to evaluate the consequences on the estuary system (Figure 3). In some instances it may be necessary to specify the structural implementation of the chosen SCDO (e.g. linear structure, groynes etc.) in order to further define the cause.

However, in most cases within an estuary the majority of defences are likely to be linear structures. The cause-consequence model aims to inform the user of the consequences of each type of SCDO, and thereby integrates processes into the resource driven approach. If the chosen SCDO has seriously detrimental implications for the estuary system, then it is necessary to go back a stage and define a new preferred SCDO.

Resource driven approaches to SCDO's usually operate over short timescales being compatible with human planning needs. Estuary wide approaches would be expected to aim at longer timescales more compatible with estuarine evolution. However, the timescale of the resource driven approach can be extended by the introduction of residual life value for the infrastructure surrounding the estuary (see Townend *et al.*, 1996). In this way infrastructure which has exceeded a certain age is effectively removed from the resource considerations. Such a timescale may well be the design life, or the period after which there has been a desired return on capital investment. By removing elements from the infrastructure, they cease to be constraints and thus widen the number of SCDO's possible. These SCDO's can then be fed into the cause-consequence model as described above.

(c)   **Final selection of SCDO's**
The final step of Stage 2, is to reconcile the chosen SCDO's from both the resource and the function driven approaches. In keeping with established SMP practice, the various options can be used as the basis for the consultation process that is an integral part of Stage 2. This consultation process needs to be approached with some care. It is likely that the SCDO's derived from the resource driven approach will favour the status quo (particularly if the analysis does not include the concept of removing life expired infrastructure). By contrast the SCDO's that accommodate a set of functional requirements may be unsuitable, particularly in the short-term. It will therefore be important to give a context to the proposed changes and stress the timescales involved.

124 COASTAL MANAGEMENT

Figure 2: Estuary Management Cause-Consequence Model

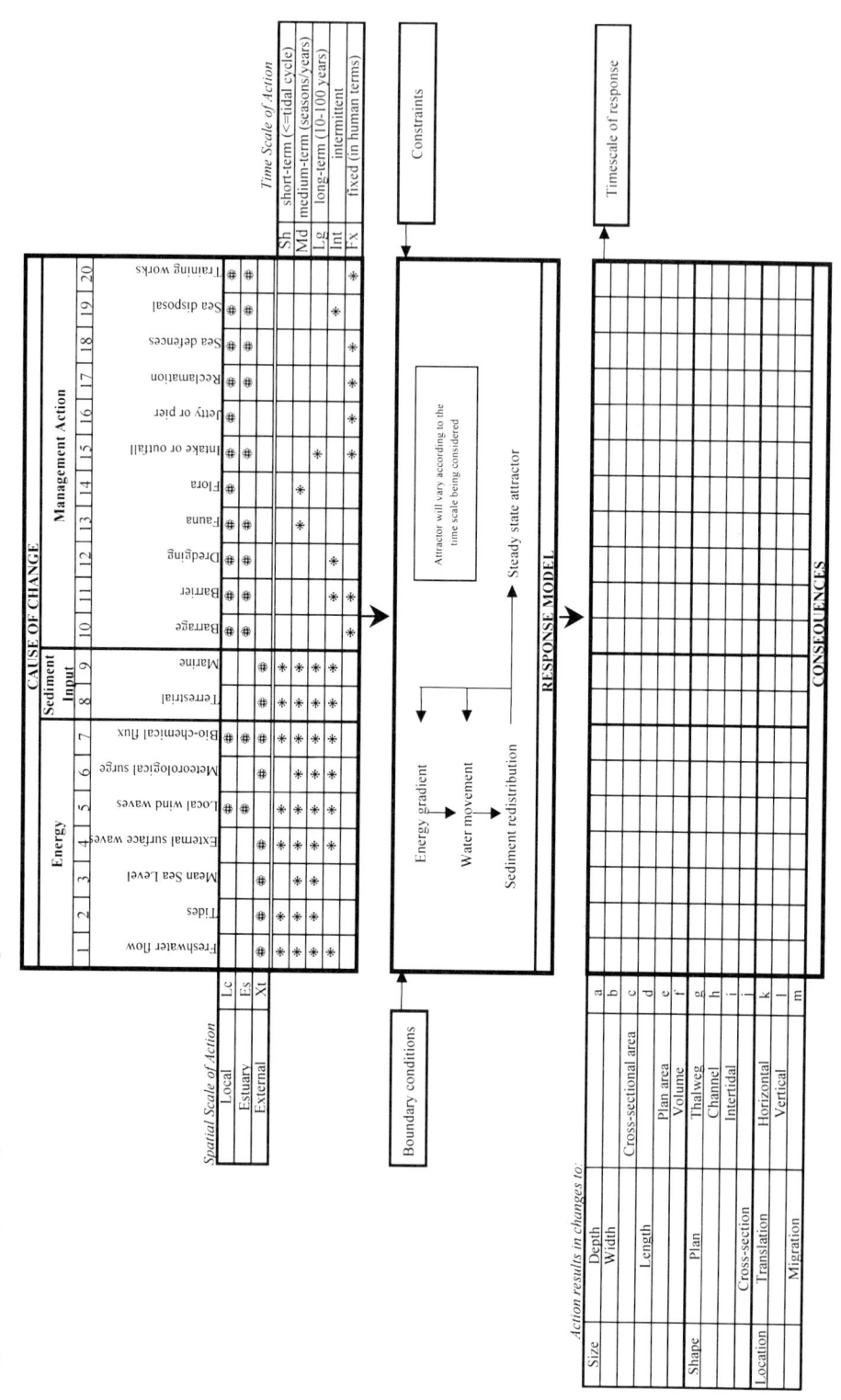

Figure 3. SCDO Cause-Consequence Model

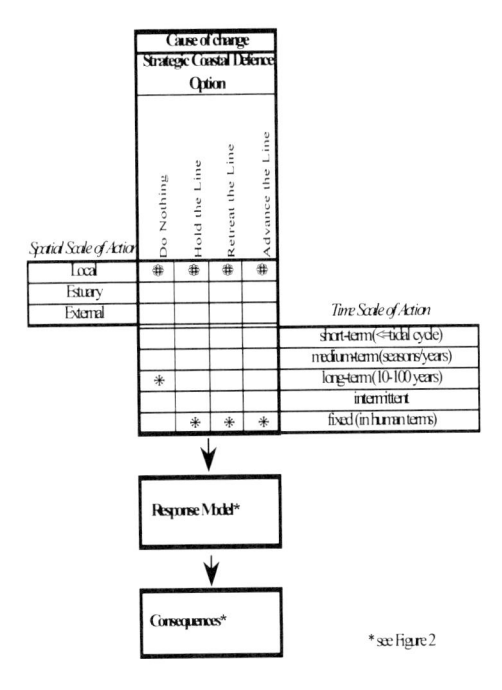

CONCLUSIONS

Shoreline Management Plans seem set to play a key role in focussing the debate upon the choice between continuing to defend the coast or accommodating change, which sooner or later will need to be made. As might be expected this issue is particularly sensitive along the more developed lengths of coast, and is especially relevant to estuaries where there are large numbers of users and conflicts. The procedure suggested for eSMP's is more extensive than that used for coastal SMP's. The steps proposed to define SCDO's are as follows:

Stage 1
(i) Identify all those with an interest in the area;
(ii) Collate and analyse existing data on all the key issues;
(iii) Set management objectives for the Plan area.

Stage 2
(i) Identify functional requirements and develop "design" concept to accommodate these functions as far as possible;
Evaluate the size, shape and location of the changes that are likely to result;
Assess the implications for **estuary resources**.
(ii) Use resource interests to define an initial set of SCDO's;
Evaluate the size, shape and location of the changes that are likely to result;

Assess implications for the **estuary system**.
(iii) Compare the options against the management objectives and consult to identify preferred SCDO's.
(iv) Map SCDO's as the management units for the estuary.

From previous experience the rule based analysis is straight forward and quick to apply (given the relevant data). The ability to apply the cause-consequence model is however dependent on the quality of the response model. An initial attempt to define this model is given in a separate paper (Pontee and Townend, 1999).

Importantly, the proposed methodology for the production of an eSMP establishes a consistent procedure for defining SCDO's, that can be revisited as the science base advances, new data becomes available, or new infrastructure needs are identified. Thus, although there will inevitably be some limits and caveats applied to the first attempt to define SCDO's, the analysis is repeatable and so the method proposed provides a sound framework on which to develop a long-term interactive management plan.

ACKNOWLEDGEMENTS
This work was funded by the Environment Agency and Ministry of Agriculture, Fisheries and Food. Nigel Pontee, Peter Whitehead and Ian Townend of ABP Research prepared the report, with valuable inputs provided by John Pethick and Jeremy Lowe of the University of Newcastle and Richard Young of Binnie, Black and Veatch.

REFERENCES
Halcrow, 1991, April 1991, The Anglian sea defence study – stage 3 – management strategy report. April 1991, Sir William Halcrow and Partners.

HR *et al.*, 1996, Estuary processes and morphology scoping study, SR446, 1-104, HR Wallingford.

Leafe, R., Pethick, J., Townend, I.H., 1998, Realizing the benefits of shoreline management, The Geographical Journal, 164, 3, 282-290.

MAFF, 1995, Shoreline management plans – a guide for coastal defence authorities. May 1995.

Pethick, J.S., 1996, The Blackwater estuary, Volume 2: geomorphogical trends 1978 to 1994. Draft report to NRA and English Nature, Cambridge Coastal Research Unit, February 1996.

Pontee, N I. and Townend, I. H. (1999). The development of a cause consequence model for an estuary system. *To be presented at:* MAFF conference of River and Coastal Engineers, 1999, Keele.

Townend, I.H., Tomlinson, B., Hill, C and Craggs, M., 1996, Defining management units for use in shoreline management planning, ICE conference – *Coastal management '95 – putting policy into practice*. Inst. Civil Enginr Conf. Proc. London: Thomas Telford: 85-99.

Montrose Bay - Coastal Management In Practice

MARTIN B MANNION, Senior Engineer, Halcrow Group Limited, Burderop Park, Swindon SN4 0QD, UK

ABSTRACT
The paper highlights aspects of coastal and shoreline management and how this is influenced by the application of science for Montrose Bay on the Scottish east coast.

BACKGROUND
Montrose Bay comprises a 9km gently curved east facing sandy beach with the River South Esk entering the bay at its southern end via a shallow muddy tidal basin, Montrose Basin (Figure 1). It is exposed to long fetches from the north-east to south-east sectors across the North Sea. The River North Esk flows into the bay towards its northern end. The town of Montrose and associated development lie in the southern half of the bay. The bay is enclosed between two rocky headlands, Scurdie Ness to the south and Milton Ness to the north. It has suffered from erosion over the years, leading to the construction of various defences at several points, although the majority of the coastline is natural, comprising a sandy beach and sand dunes.

Concern about coastal erosion led Angus Council and GlaxoWellcome to jointly commission Halcrow to undertake a Shoreline Management Study of the bay. Scottish Natural Heritage and Montrose Port Authority were involved throughout the study. The study focused on two areas of concern and aimed to develop appropriate solutions, within the context of understanding the coastal processes of the bay and its environmental setting. The natural assets of the area are reflected in the designations assigned to the area. These include Montrose Basin, a Ramsar site, and St Cyrus, a National Nature Reserve. The study followed MAFF guidelines (ref 1).

DEVELOPMENT OF EXISTING DEFENCES
Records and information on the existing defences were collected, a detailed site inspection undertaken and data analysed and stored in Halcrow's SANDS (Shoreline And Nearshore Data System). The defences have been extended in stages over time.

The GlaxoWellcome pharmaceutical site is bounded to the south by the River South Esk, to the east by the beach and to the north by a caravan park. It employs some 700 staff in a complex of 160 buildings with a replacement value of £100 million. It has gradually developed seaward over the last few decades to occupy an area of former sand dunes. From 1970, gabion protection was provided to the seaward edge. Deterioration led to further rock armour protection over a 600m length in 1990/91.

The caravan park to the north gradually suffered increasing erosion of the sand dunes, leading to construction of a 350m long rock armour revetment in 1991. Immediately to the north of the caravan park is the Pavilion seawall. The 200m long seawall comprises a slope of pitched

Coastal Management: Integrating science, engineering and management, Thomas Telford, London, 2000

128 COASTAL MANAGEMENT

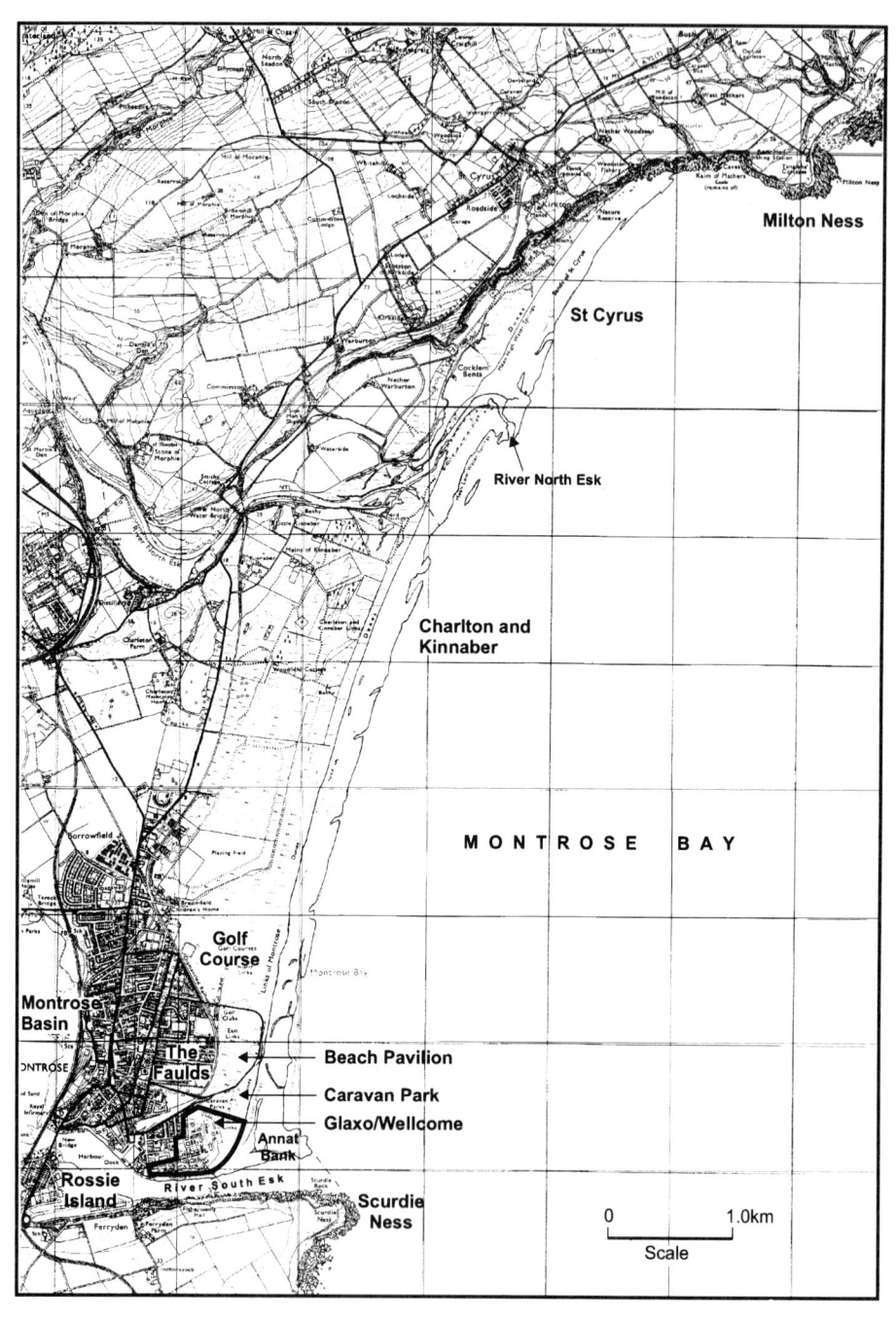

Figure1 MONTROSE BAY LOCATION PLAN

stone in asphalt and wave return wall. It was built in 1954, replacing an earlier less substantial structure. Falling beach levels led to increased storm damage and the addition of a rock toe in 1989/1990, and terminal erosion protection at the northern end in 1991. The Pavilion seawall is further seaward than the defences on either side.

To the north of the seawall, there is a step back in the coastline due to erosion and the golf course frontage extends northwards towards the middle of the bay. It comprises some 15m high sand dunes, with protection limited to three number rock strongpoints providing limited protection to three tees on top of the dunes.

North of the golf course, the coast is undeveloped and the sand dunes gradually decrease in height until the River North Esk. Here a north-east oriented spit diverts the River North Esk to flow out in front of the extensive dunes at St Cyrus. The dunes become narrower to the north, before meeting the rocky cliffs of Milton Ness.

ENVIRONMENTAL ISSUES AND CONSULTATION
The continued erosion of the dunes and beach has led to problems affecting both the leisure related activities of the links including the golf course, and the Glaxo Wellcome site itself. The beach in the southern half of the bay is easily accessible by the public and utilised by locals and visitors. Access is limited between the golf courses and River North Esk whilst St Cyrus has a visitor centre and is a popular nature and leisure site.

Much of the area is of very high nature conservation value, containing two designated sites, Montrose Basin SSSI and St Cyrus SSSI and NNR. Montrose Basin is a large estuarine basin of the River South Esk, exposing extensive mudflats at low tide, designated for its coastal habitat, important to overwintering wildfowl and waders and for geological stratigraphical exposures. St Cyrus is one of the richest and most important sites for flora and fauna on the north east coast of Scotland. There is occasional salmon fishing on the foreshore, particularly near the North Esk.

The Glaxo Wellcome site is a major piece of industrial infrastructure, contrasting with open landscape amenity areas to the north. The Pavilion frontage has been the subject of recent extensive environmental improvements to the leisure facilities. The extensive links area to the north is given over to two 18 hole golf courses. The outer Medal course is reputed to be the 5th oldest course in the world. Three tees are at great risk from erosion, having already been relocated inland from former positions. The Pavilion area and golf courses are separated from the town by open amenity land. North of the golf courses, rough pasture and small coniferous and deciduous tree plantations extend as far as the River North Esk, behind the dunes.

COASTAL PROCESSES STUDIES
The coastal processes assessment included historical coastal evolution, wave modelling, beach modelling and tidal analysis. The bay comprises glacial deposits that have been reworked to create the modern landforms, by sea level (changes), river, wind and wave processes. There is sedimentary evidence (ref 2) for a historic tsunami some 7000 years ago, which had a height of at least 3m above sea level. The spring tidal range is 4.2m, with the 1 in 100 year water level being identified as 1.42m higher than MHWS.

The sandy foreshore has shore parallel sand bars which migrate onshore seasonally, their form altering near the two rivers. The dunes in the southern half of the golf course are subject to ongoing undercutting by wave erosion, whilst to the north there is evidence of episodic storm erosion. At St Cyrus, there are currently embryo foredunes although locally erosion and accretion is influenced by periodic growth of the spit across the mouth of the River North Esk. Growth causes consequential erosion to the north, as the river is pushed towards the dunes, until eventual breach and roll ashore of the spit. Records of spit extent and age indicate that this cycle is typically 15 years (ref 3).

The River South Esk has been dredged over the last 25 years, to maintain the channel for navigation access to Montrose Port. Sand and some cobbles are dredged and dumped in a licenced site offshore of Lunan Bay (south of Scurdie Ness). Typically 51,300m3 has been dredged per annum, removing it from the local sediment system.

Comparison of former high water and low water positions with the present shows that the rate of erosion has been greater within the last 20 to 30 years compared to over the last 130 years. The area of erosion has predominantly been in the southern third of the bay, from the middle of the golf course frontage southwards to the River South Esk. The principal erosion is by wave action except towards the River South Esk where tidal flow plays its part in sediment movement. Wind transport is of less significance.

Angus Council had undertaken biannual beach profile surveys in the southern third of the bay for several years (1989 to 1996). This data was to prove extremely valuable in understanding the coastal processes within the bay.

The data was analysed in SANDS to identify mean annual shoreline movement, beach level changes and volumetric changes (Figure 2). For the GlaxoWellcome frontage, average beach lowering was up to 0.4m per year for some profiles, whilst the caravan park frontage to the north was stable. Meanwhile the Pavilion Seawall frontage foreshore was lowering by 0.3m per year, but with considerable fluctuations. The southern half of the golf course also suffered erosion, with a 5m average MHWS recession and 0.2m beach level fall. Towards the middle of the golf course frontage, erosion decreased. The long term rate of volume loss was 51,000m3/year.

Five years of offshore Met Office data (1992 to 1996) for 56.75 N, 1.66 W was purchased. The time series data includes wave height direction and period data, wind wave and swell wave components and wind speed and direction. Wind data from Leuchars has been previously analysed (ref 4) for the years 1970 to 1991, as well as Bell Rock and Fraserburgh data. Comparison of the 1970 to 1991 data with 1987 to 1991 showed that onshore winds had reduced particularly between 015 and 175 N.

Further comparison with wind data from 1992 to 1996 showed that the percentage of winds is normally greater from 120 to 180 N than the 30 to 90 N sectors. Analysis of swell and wind waves showed that wind waves were predominantly from the south east whilst swell waves were from the north east.

Wave transformation modelling was utilised to provide a time series of nearshore wave conditions, utilising Halcrow's MWAVE grid based model, taking account of wave diffraction, refraction, breaking and seabed friction (ref 5). A 20m grid was used nearshore, nested in a 75m grid offshore. Inshore wave climate for a 1 in 100 year storm (Hs 4.7m, Tz 7.8 to 10.3s) was used for design. Nearshore wave climate, beach grain sizes and profiles were utilised in Halcrow's Beach Plan Shape Model (BPSM).

PAPER 13 : MANNION *131*

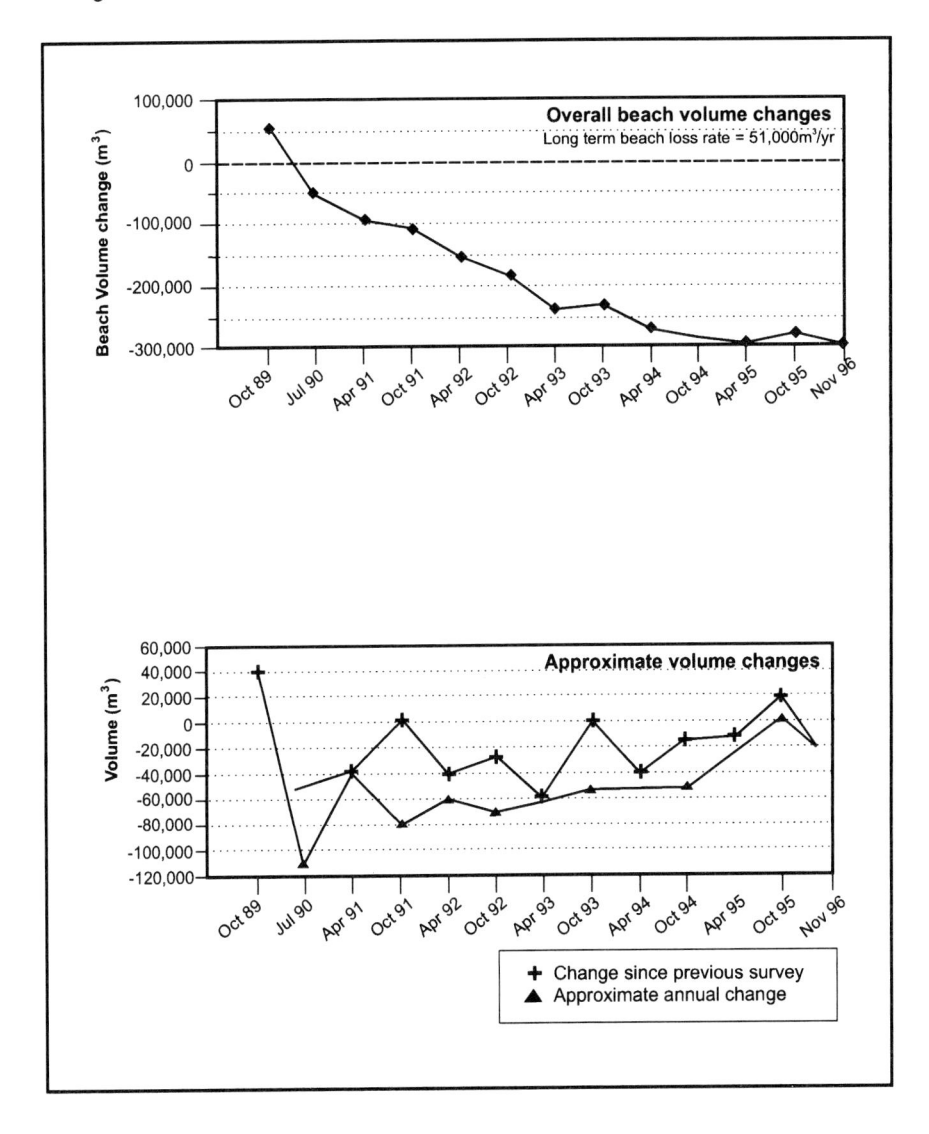

Figure 2 **BEACH PROFILE ANALYSIS - VOLUME CHANGES**

BPSM provided model predictions for beach evolution, for the existing situation and for coastal defence options. BPSM is best suited to the study of long term changes in beach plan position and volume changes. It was calibrated using the available beach profile data, with year on year changes being directly linked to year on year wave climate, to assess variability of beach response. Calibration against beach profile data, combined with analysis of earlier hydrodynamic modelling (ref 4) ensured that the results were representative of the combined influence of tides and waves near the River South Esk.

BPSM results showed that there was a net loss from the system due to southerly drift from the southern third of the bay of 20,000m3 to 80,000m3/year into the River South Esk. The average loss of 49,500m3 compared well with the average dredged volume (51,300m3). Figure 3 shows actual dredged quantities for individual years against annual results from BPSM. The northern part of the bay was in dynamic equilibrium, with limited accretion at the River North Esk.

Resultant beach change was found to be heavily linked to wind waves in the northern bay and swell waves in the southern bay. Increases in south-easterly winds lead to increased overall erosion. This trend of increased easterly winds over the last 10 years or so (compared to the last 50 years) is predicted to continue (ref 6). Thus use of 1992 – 1996 wave data and 1989 – 1996 beach profile data is suitable for predictions of future coastal charge, to inform shoreline management.

Beach profile response to storm action was modelled with COSMOS -2D, in order to understand the extent of beach drawdown, bar formation and how this would effect the design of beach control structures.

COASTAL DEFENCE OPTIONS

In line with MAFF recommendations (ref 7), a wide range of options were considered under the following categories, for a 50 year design period:

- Do Nothing – take no action to maintain or improve existing defences (baseline)
- Risk Management – improve warning systems to safeguard lives
- Sustain – carry out works to maintain current standard and line of defence
- Change – improve standard of defence

Extensive consultation was undertaken with statutory bodies and the public, including public meetings and exhibitions, brochures and use of the press.

GLAXOWELLCOME FRONTAGE

For the GlaxoWellcome Site, there was a clear economic benefit for appropriate sustainable defence works. The following options were considered.

Option	Effect
Do Nothing	Failure of existing defences within 5 years, loss of site and jobs
Maintain existing defences	High cost, risk of failure, beach lowering continues
Beach recharge	High cost, increase in dredging, disruption to shipping
Rock/concrete revetment	Large high cost structure, continuing beach lowering, visually intrusive

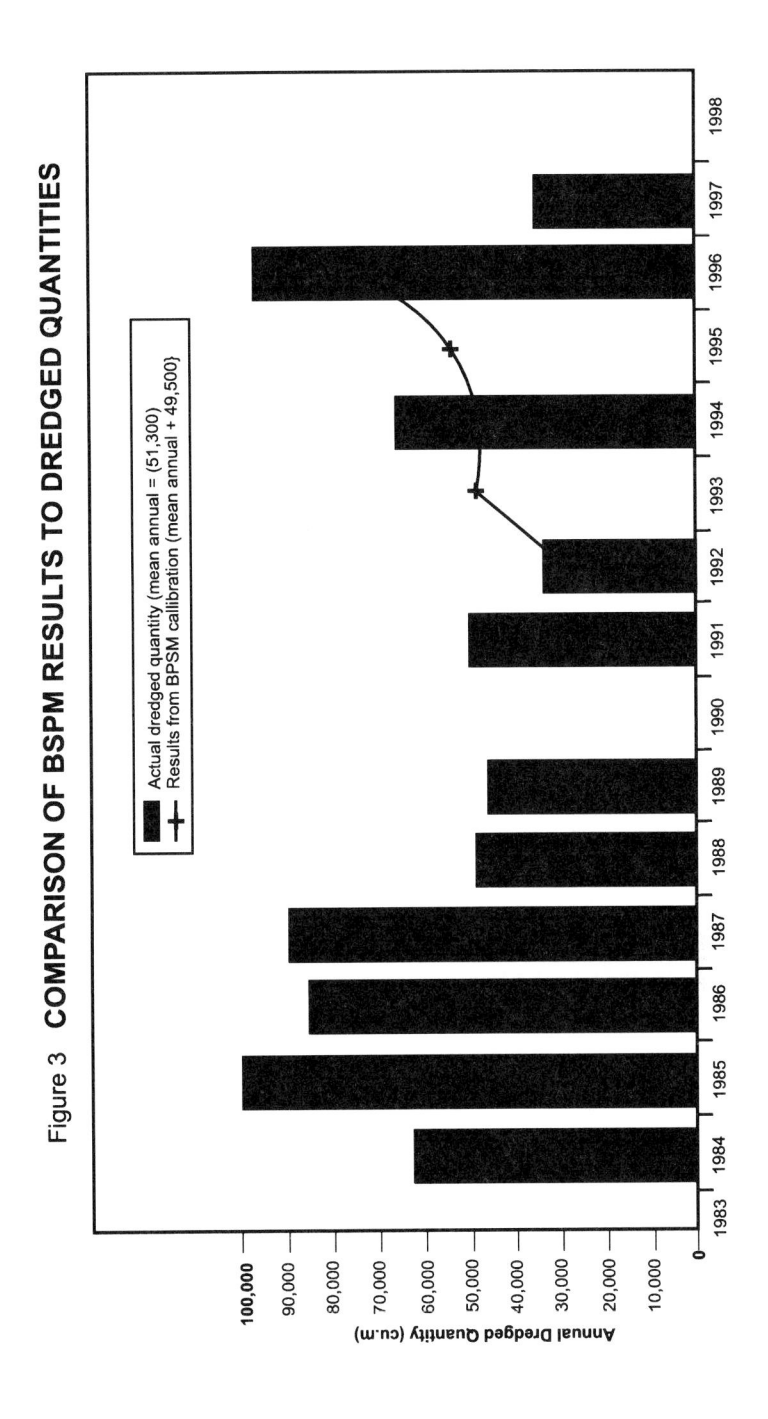

Figure 3 **COMPARISON OF BSPM RESULTS TO DREDGED QUANTITIES**

134 COASTAL MANAGEMENT

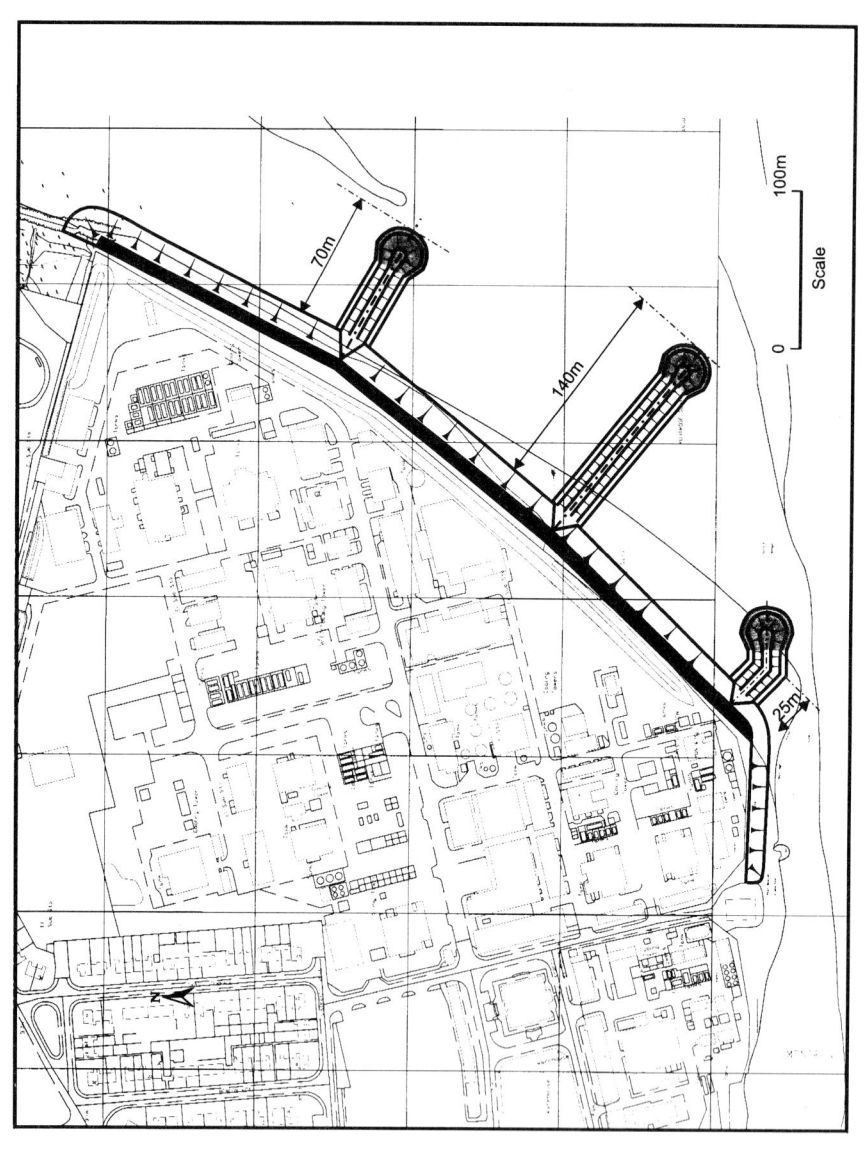

Figure 4 **GROYNES LAYOUT PLAN**

Vertical seawalls	Increased beach scour, high cost, visually intrusive
Offshore breakwater	Shipping hazard, high cost, visually intrusive
Groynes and beach recharge	Medium cost, some visual intrusion, beach retained

The groynes and beach recharge option was recommended as it would provide long term protection with minimal impact, downdrift dredging of the River South Esk would reduce and use of the beach would be retained. It presented the best overall economic, technical performance and environmental option.

The option was developed further through statutory, community and staff consultation. Sea supply of the principal materials, rock armour and sand/shingle, was specified. Recharge material will be obtained from the River South Esk where beach material drifts under sediment transport. Construction began in May 1999 and will be completed by October 1999. The scheme involves importing 50,000 tonnes of 1 to 8 tonne rock armour and 70,000 m^3 beach recharge, the construction of 3 rock groynes of 50m, 140m and 70m length, upgrading the existing revetment (re-using rock where possible) and raising beach levels by up to 2 metres (Figure 4). It is estimated that beach top-ups would be needed typically every 20 years.

Local involvement such as schools poster competitions, information handouts and questionnaires, joint preparation of schools educational material and production of a video will assist in ensuring that ownership is taken for future coastal management.

GOLF COURSE FRONTAGE

For the golf course frontage, similar options were investigated. For 'do nothing', up to 80m of erosion was predicted within 50 years, leading to the loss of four holes. Revetment and groyne options were expensive with visual impact and increase erosion on the adjacent coastline. Partial intervention was also considered, reinforcing the existing ineffective rock strongpoints. Beach recharge was too costly in benefit-cost terms. Breakwaters were not an effective technical/economic solution.

A variation on 'do nothing' was considered, namely landward intervention and 'do nothing' to the shoreline. This involved recreating the 4 holes due to be lost inland on unutilised open land. The course would also need to be altered. Relocation was estimated to cost £740,000 compared to benefits (potential losses) of £2,120,000 and the cheapest intervention option (bolstering the strongpoints would cost £1,850,000 with reduced benefit (losses) of £350,000).

Given the historic heritage of the golf course and its value to the community in leisure and tourism terms, there was a strong lobby for intervention. The course will be a (British) Open qualification course in summer 1999. Following extensive consultation, the do nothing and relocation landward option was agreed.

REST OF BAY

For other areas within the bay, the following recommendations were made :
- Caravan park frontage - Hold the line, no works in the medium term.
- Pavilion Seawall - Hold the line in the short term. Works likely to be required to prevent collapse. Beach recharge and landward realignment to be considered.
- North of golf course/remainder of bay - Do nothing

It was recommended that future dumping of maintenance dredged material be near low water offshore of the Pavilion Seawall, to reduce erosion and retain sediment in the bay. Discussions with the port and fisheries interests indicate that this beneficial use will become practice. The existing monitoring has also continued and been extended.

CONCLUSIONS

Particular findings were that long term predictions are crucial to such finely balanced sediment systems and that climate change is altering the beach dynamics by increasing erosion patterns, due to changes in wind climate. Beneficial use of dredged material will also assist coastal managers in adopting to coastal change. Acceptance of beneficial use involved discussion and dissemination of the advantages with all parties.

The successful undertaking of the study relied on sufficient relevant long term data and the application of coastal process modelling to achieve understanding of the past and future evolution of the bay. In this case, regular beach profiles by Angus Council were particularly useful for the wave and beach modelling elements of the study.

At one focus area (GlaxoWellcome), engineering solutions were of greater priority although the ultimate solution achieved a successful balance between engineering and environment, whilst for the other one, environmental considerations were of more significance. For the GlaxoWellcome site, works are now under construction.

At the other focus area (the golf course), a policy of "do nothing" was recommended for the coastline, with the relocation landward of the portion of the golf course predicted to erode. This option involved balancing recreational, environment and socio-economic considerations. It represents coastal rather than shoreline management.

Although no Shoreline Management Plan is in place for the area, the study mirrors the approach taken to carry out Strategy Studies in England. It involved extensive consultation and benefited from the co-operative approach of Angus Council, GlaxoWellcome, Scottish Natural Heritage and Montrose Port Authority.

Early involvement of the public and key interested parties in decision-making on the coast is crucial to achieving acceptance and ownership of sustainable shoreline management policy and implementation details.

ACKNOWLEDGEMENTS
The author wishes to thank GlaxoWellcome, Angus Council for permission to publish the paper and colleagues at Halcrow, MAFF and SNH for comments.

REFERENCES
1. MAFF (1993) 'Coastal Defence and the Environment, A Guide to Good Practice'
2. Dawson, A.G. et al (1988) 'The Storegga Slides : Evidence from Eastern Scotland for a possible tsunami' *Marine Geology*, 82. 271-276
3. Croft, R.S. (1976) 'Landform Development, River North Esk'
4. HR Wallingford (1995) 'Montrose Coastal Erosion Study Phase III, Tidal modelling of Montrose and Lunan Bays' Report EX3195
5. Bin, L & Fleming, C.A. (1997) 'A three dimensional multigrid model for fully nonlinear water waves' Coastal Engineering 30 (1997) 235-258
6. HR Wallingford (1991) 'Wave Climate Chane and its Impact on UK Coastal Management' Report SR260
7. MAFF (1993) 'Flood and Coastal Defence - Project Appraisal Guidance Notes'

The Legal and Geomorphic Impacts of Engineering Decisions on Integrated Coastal Management

D. J. MCGLASHAN[1] and G. R. FISHER[2]

[1] Department of Geography and Topographic Science, University of Glasgow, Glasgow, G12 8QQ, UK. *Present Address:* Graduate School of Environmental Studies, Wolfson Building, University of Strathclyde, Glasgow, G4 0EU, UK.

[2] Tutor, Law School, University of Glasgow, Glasgow, G12 8QQ, UK

INTRODUCTION

This article is written for coastal managers, and coastal actors of all kinds. The question it poses is what legal issues *should* coastal actors bear in mind when taking their decisions when planning at the coast? This debate can take place on two levels, and both are discussed here. Firstly and practically, what issues must be considered on a day-to-day basis at present. In this we focus particularly on the direct application of basic legal principles to coastal erosion specifically.[1] At this first level, emphasis is laid not on the basic statutory functions that empower local authorities and other bodies at the coast, but rather on the legal duties that can arise to bind these bodies and other coastal actors. Secondly, an evaluative examination of the long-term consequences of this approach is presented from the perspective of environmental science. Throughout, the common law system of *ad hoc* resolution of disputes between private interests is assessed in comparison to a more planned scheme, for example the model of the revised Environmental Assessment regulations. The interplay of the two levels of debate will be seen as central in the legal resolution of the issues.

COMMON LAW LIABILITY AT THE COAST

These issues are typically very emotive, and thus frequently involve litigious, as well as politically active, landowners and tenants. The merits of any claim aside, the necessity of legal advice and Counsel's opinions in complex areas of law can be prohibitively expensive, although these problems may be helped in the long-term by the recent reforms of the English legal system.

The various actors in any imagined coastal defence scenario can, quite apart from their legal obligations in Government legislation, become subject to legal duties giving rise to liability or a duty to act. These duties (common law) regulate the competing private interests at stake, and can apply to landowners, operating authorities, tenants, developers and coastal experts. Recent years have also seen an increase in the litigiousness of those with an interest near the coast. The common law is still useful for various reasons, despite the utility of the statutory provisions noted below.

It should be stressed that the duties discussed here, although sound in scientific theory, have not

---

[1] The legal position referred to is principally that in England and Wales, as this will affect most actors at eroding coasts.

yet been seen in practice to a great extent in the courts, no doubt to the relief of coastal managers everywhere. The fact remains that legal advisors in many cases would be forced to caution their clients on the risks that could arise where the right conditions and facts prevail, as the liability results from the application of known and general principles of the law.

The practical legal examples that have reached the courts immediately focus the attention on the question of liability. The *Holbeck Hall Hotel* case will be first in most readers minds.[2] That case highlights one possibility for liability at the coast. The scientific and engineering aspects of the case for legal liability do not address directly the mechanics of the coastal system, but depend rather on subsidence and landslip, albeit arising ultimately from a collapsing coast.[3] The case acts to confirm and strengthen the principle that natural processes arising on land can still source liability of the land occupier, following the reasoning used in *Leakey v National Trust*. This has consequences for public authorities in particular, as the standard of care depends in these cases on the resources of the defendant, as it is termed a 'measured duty of care', and has a knock-on effect of strengthening the type of claim mentioned here as a second possibility. The Holbeck Hall Hotel case may well however be shortly overruled, as there is an appeal to the Court of Appeal pending.

The second possibility - and the one which centrally involves coastal geomorphology and this article in particular - is where coastal processes directly create a claim. Here as yet there have been no clear cases. The facts of the case are always crucial. We envisage a case where landowner A owns a stretch of coastal land adjacent to the coastal land of landowner B. A's land is immediately up-drift of B's land. Both properties are situated on a sedimentary coast within the same coastal cell. In such a case the situation might arise where B has a possibility of legal action against A because A has put in place coastal defences which affect B's land.

The legal basis envisaged for such a challenge is primarily the law of tort or delict, where harm is caused by the legally culpable action of another, through a challenge likely to be founded on the more specific ground of nuisance.[4] The principal remedy sought will be one of injunction or interdict to prevent the continuance of the nuisance, although damages may also be available, and this is discussed in detail below.

In any such case there will be no possibility of legal action unless B can point to both a loss or injury that his land has suffered, and a process that - on balance of probabilities - gave rise to that loss or injury. It is here that an understanding of the geomorphology is critical.

GEOMORPHIC PROCESSES

Given the scenario above it is worth covering the important geomorphic issues. The coastal system is in a constant state of change. The morphology of the coast is controlled by the interaction of energy and materials within the geologic framework (Hansom, 1988). The coast operates as a transport system, with sediment as the cargo, which is eroded from source areas and deposited in sink areas (there is limited scope for reviewing this information in depth; the reader

---

[2] *Holbeck Hall Hotel Ltd, & anr. v Scarborough Borough Council & anr.*, 1997 2 E.G.L.R. 213.

[3] *Ibid.*, per Hicks, J. in para. 5 at 214D, and more generally; see also Fleming (1992: 11)

[4] Another corollary of the *Leakey* case is that it confirms it is not necessary to specify a specific tort, provided the facts of the case disclose such a tort.

is referred to Carter, 1988; Hansom, 1988; Komar, 1998). As the processes of erosion, transport and deposition operate, longshore sediment transport directions can be identified, and once these are known coastal cells can be mapped. The entire UK coast has been split into cells and sub-cells of self contained sediment circulation (HR Wallingford, 1997; Motyka & Brampton, 1993). Each coastal cell will have both source and sink areas and the transport between them should not pass outwith the boundaries of the cell. When undertaking any work at the coast it is imperative that the sources, sinks and transport routes of sediment are known. A sediment budget should be produced which at the very least considers the credits and debits to the system. The credits to the coastal system come from coastal erosion and river input, though some may come from offshore. The debits are in the form of losses inland of fine particles (sand, silt etc.) and losses offshore. However, human interference has added to the number of debits in the coastal system through sediment trapping in dams and shore normal structures, reduction of sediment input by coastal protection, and marine sediment extraction. These human interventions can then result in loss suffered by other coastal actors, as follows.

IMPACTS OF PROTECTION STRUCTURES

There are two main types of shore protection structure: shore parallel and shore perpendicular. Shore parallel structures tend to have a number of effects, the severity of which will vary depending on the extent, style and construction of the structure. Within the immediate area of the structure there is often evidence of basal scour at the toe of the structure, beach draw-down (the lowering of the level of the beach and reduction of the width of the intertidal zone), overtopping (where material is eroded from behind the structure by overwash) and flanking erosion (where erosion is focused at the ends of structures) often producing an erosional bight (for reviews of the effects of coastal structures see: Kraus & Pilkey, 1988; Tait & Griggs, 1990; Bird, 1996; Komar, 1996; Kraus & MacDougal, 1996). The majority of these can be reduced by careful design; whether they can be eliminated or not is subject to debate. However, arguably the greatest loss to the down-drift part of the cell is the reduction of sediment supply. This reduction of sediment will have an adverse impact on the down-drift section of the cell by either compounding existing erosional problems, or inducing new ones. Shore normal protection structures are designed specifically to reduce volumes of sediment transported down-drift. These are typically in the form of Groynes which are designed to 'hold' sediment on the up-drift side of the structure. Groynes are often placed to combat beach draw-down resulting from shore parallel protection structures.

In order to link the loss to the act or omission of the defendant complained of, the law will analyse whether it actually caused the loss. This is a two question process: (1) whether the loss would have occurred without the act or omission; and (2) whether the act or omission is the legally significant cause. The first question is legally simple, and is a question of fact for expert testimony. Where, as here, there are multiple causes, it is sufficient that the defendant's act is a material contribution to the loss. The second question is complex, and is resolved by the courts as a matter of law; it is dependent upon a 'common sense' approach, from the point of view of the ordinary man, rather than that of the scientist or the metaphysician.[5] The vital test here is whether the defendant could reasonably foresee the consequences of their action. With basic coastal processes and the effects of coastal structures being documented since at least the earlier half of the century (Mathews, 1934), foreseeability should not necessarily be a bar to liablility, although ultimately it will depend on the circumstances of the individual case.

---

5 *Yorkshire Dale Steamship Co. Ltd. v Minister of War Transport* [1947] A.C. 691, 706.

LEGAL THEORY

In the type of case we envisage, with relatively 'direct' damage to the private property rights in a plaintiff's land, private nuisance is the most likely ground.[6] Private nuisance is, to paraphrase, unlawful or undue interference with the use or enjoyment of land or some right in relation to that land. One of the forms of nuisance is direct physical injury to land,[7] although damage to land itself does not render the act or omission a nuisance. The damage must also be continuing or repeated. There will then be a complex weighting of the interests involved, to determine whether the nuisance is actionable or preventable and a variety of circumstances are taken into account in this process, which essentially becomes - as so often in the law - one of reasonableness and fact and degree.

Given the lack of direct authorities immediately in point, the question arises whether nuisance extends to our situation involving A and B. In this context it is worth noting the question addressed by Hicks, J. in the *Holbeck Hall Hotel* case, after mentioning the test of whether preventative measures would have been taken by a reasonable person in the defendant's position:-

> One way of framing the question posed by the present claim is whether that test should be extended to situations in which the potential harm to the neighbour's enjoyment of possession arises from failure of support rather than encroachment of materials or conditions.

The question was answered in the affirmative in the case. To establish liability, reliance was put on the relatively novel concept in the law of a 'measured duty of care' which results when a hazard arises on land which then adversely affects another's land. Our situation envisages the defendant taking positive action which results in harm to the plaintiff's land, so the liability for failure to act element of the *Holbeck Hall Hotel* case may not be necessary for liability to arise, but the result of the case does confirm that nuisance applies to this kind of situation.

Further problems arise, firstly from a line of Nineteenth Century cases as discussed in *Langbrook Properties Ltd. v Surrey County Council* following on from the House of Lords case of *Bradford v Pickles*, that there is no interest at common law in percolating or underground water until it is appropriated.[8] In the *Langbrook* case the authorities were reviewed to the effect that a landowner has no absolute right to water support as he does to support by land, although it has been observed that in none of these cases had the defendant caused actual damage to land by negligent abstraction of water (Weir, 1996: 445). It is suggested that this particular class of case, relating specifically to the abstraction of percolating water, cannot be extended to our situation, where it is not the interest in the water that is relevant but rather the interest in land (sediment). This will also be a case of such damage to land actually caused.[9]

---

[6] A nuisance can be classified a public nuisance if it affects a class of the public, and can still be actionable if the coastal actor can show they have suffered special damage.

[7] It seems clear that it need not involve some kind of physical emanation; *Hunter v Canary Wharf Ltd.* 1997 2 W.L.R. 684

[8] *Langbrook Properties Ltd. v Surrey County Council* [1970] 1 W.L.R. 161; *Stephens v Anglian Water Authority* [1987] 1 W.L.R. 1381; *Bradford v Pickles* [1895] A.C. 587.

[9] See also *Brace v S.E. Regional Housing Association* [1984] C.A.T. 20, for a contrary point of view.

Secondly where the use of the land alleged to be a nuisance is an ordinary or normal use of the land, it is protected and no liability arises. Here we imagine the nuisance to result from a perfectly legitimate use of land that is likely - in isolation at least - to be of benefit to the community. This would indicate that damage should be borne by the affected proprietor. This is only true, however, up to the point that serious damage is caused to the plaintiff's property. The building of a retaining wall has been held to be a normal use of land.[10] However in these cases the nuisances were unknown to the defendants, as in one case the damaged sewer was underground on the defendant's land and so unknown to them, and in the other the defendant did not know of any defect in the wall before it collapsed, causing damage. This fact, and the development of the law since, particularly in the principle arising from the *Leakey* case, may well leave an opening for liability at the coast to arise.

There may exist a natural right to the support of land in its natural state (typically the state we are faced with in our example), or an easement or servitude right to support land or buildings. In such a case the conduct of the defendant, the nature of the locality and the benefit of the activity to the community will not matter, as they would in a typical nuisance situation. This does however depend greatly on the individual circumstances at the point of the coast in question.[11] The principle on percolating water may well affect the application of this grand challenge.

In the weighing process the conduct of the defendant is usually relevant, although the fact that a defendant has taken all reasonable care to reduce the nuisance is not ultimately a defence, if the reduced nuisance is still deemed by the law to be just that. As noted, coastal erosion is physical damage to land, and even if there is no natural right of support or easement that can be relied on for a more absolute case, physical damage falls into the category of harm where there is 'sensible injury to the value of the property' or 'material injury, and this at least has the effect of rendering the nature of the locality of no importance.'[12]

An owner or tenant may be able to sue, depending on whether they are affected by the nuisance. It is hard to imagine a situation where an owner would not be able to sue, given the nature of the damage incurred. The nuisance must in any case be significant, and it must be more than temporary in duration for an injunction to be granted.[13] If damages are sought, duration is only one factor among many in determining if the purported nuisance is reasonable. If averments of negligence are required for damages, it is notable that again the type of damage aids the case: a duty of care is relatively easy to establish where there is a claim for physical damage.[14]

When assessing the question of liability for damages rather than simply injunction for the cessation of the nuisance, it is relevant whether the defendant in any case has created the nuisance or merely continued it. There is ultimately no question of liability where the damage

---

10 *Ilford Urban District Council v Beal and Judd* [1925] 1 K.B. 671, 1925 All E.R. Rep. 361; *St. Anne's Well Brewery Co. V Roberts* [1928] All E.R. Rep. 28.

11 Another interesting possibility is the argument that a down-drift proprietor has a profit a prendre to take the sediment; *Blewett v Tregonning* (1835) 3 Ad & El 554, involving the right to take sand from another's land.

12 *St. Helen's Smelting Co. v Tipping* (1865) 11 HL 642; *Halsey v Esso Petroleum Co* [1961] 1 W.L.R. 683.

13 Liability is also limited by the fact that 20 years prescription can legalise a nuisance.

14 *Caparo Industries plc v Dickman* [1990] 2 A.C. 605, per Lord Oliver.

could not possibly have been foreseen. This is particularly relevant to this type of case - it can be seen that the state of awareness of the engineering profession and coastal experts of all kinds can itself create liability, as well as going to the question of foreseeability in causation, as noted above.[15] The *Cambridge Water* case makes it clear that no liability can arise where a member of a profession would not have been aware of the effects at the time.[16]

The question of who is liable is also important. The creator of the nuisance may first of all be liable. The occupier is then generally liable, and if their servant causes the nuisance, they are liable as a matter vicarious liability. A principal will not usually be responsible for an act of an independent contractor, but may be if the occupier has a non-delegable duty of care, as is often the case - where there is a special danger of nuisance occurring. The case may well depend on whether there is negligence on the part of the contractor, which causes the damage. An occupier will not be liable for damage resulting from an 'act of nature' unless he uses the state of affairs for his own purpose or continues the nuisance - this last is where he has actual or assumed knowledge of its existence and he fails to take reasonably prompt and effective moves to remove it.[17] A landlord may well still be liable if he has authorised the nuisance or he knew or ought to have known about it.

The defences of Act of God and the 'secret unobservable forces of nature' can lie. Act of God essentially depends, however, on the event being unforeseeable, and this is very often not the case here. Increasingly the courts take into account the increased state of knowledge. Certainly extraordinary winds and tides have been held not to be Acts of God.[18] Coastal erosion is, of course, not unobservable.

THE POWERS OF PUBLIC AUTHORITIES AT THE COAST

Overall responsibility for planned coastal defence in England is vested in MAFF, which provides guidance and funding aid for the operating authorities that take the initiative on any particular project. These 'Maritime' district councils have the principal legal powers and responsibilities in the area of coastal defence, abetted by the voluntary local defence groups, funding from the County Councils, and with a duty to consult with the Environment Agency under its aegis of responsibility for co-ordinating flood defence. The statutory and other public authorities can, however, equally fall under the common law duties discussed above, although the law does take account of their special position in various ways.[19]

Local authorities also have a duty to act where they determine a statutory nuisance to exist. This is defined under the Environmental Protection Act,[20] and can often aid a private coastal actor in

---

[15] In Scotland the recent case of *Kennedy v Glenbelle Ltd.* 1996 SLT 1186 allows that culpability may arise through recklessness (Thompson, 1997).

[16] *Cambridge Water Co v Eastern Counties Leather plc* [1994] 2 A.C. 264

[17] *Sedleigh-Denfield v O'Callaghan* [1940] A.C. 880; *Goldman v Hargrave* [1967] 1 A.C. 645.

[18] *Cushing v Walker & Sons* [1941] 2 All E.R. 693; *Greenwood Tileries Ltd. v Clapson* [1937] 1 All E.R. 765.

[19] This is by way of a defence of statutory authority, and greater latitude in the common law analysis of whether a duty of care in negligence arises, although it may not be relevant where the authority is sued simply as an occupier of land.

[20] Environmental Protection Act 1990, Part III. See s. 79, as amended, for the definition of statutory

dealing with a problem, by way of a complaint to the authority. Private coastal actors can also use the Act to bring nuisance proceedings, where the authority decides to take no action.[21] The issue here, in the type of proceedings we suggest, is whether the erosion can be termed a nuisance under s. 79(1)(a) of the Act, *viz.* whether it renders "premises in such a state as to be prejudicial to health or a nuisance". 'Premises' are specifically defined so as to include land, although there may be problems in extending the definition to a situation where there are unlikely to be public health or human comfort considerations, except in extreme cases.[22] Whether or not the nuisance constitutes a nuisance at common law can also be a determining factor here.[23]

Whether the authorities have to act themselves to enforce the anti-nuisance provisions, they may be subject to the duties as coastal proprietors. This can conflict with their other obligations, particularly with the use of their power under the Coast Protection Act 1949 to protect the coast. The case-by-case and inherently unpredictable common law can cause short-sighted action and consequences that extend beyond the locale on which they depend. The whole scheme of the policy basis on which the authorities should take decisions, now following the shoreline management plan (SMP) system is that of rational planning.

ROLE OF THE LAW WITHIN INTEGRATED COASTAL MANAGEMENT

Despite the best efforts of MAFF, English Nature, the Environment Agency, Scottish Natural Heritage and numerous academics, private coastal actors tend to respond to erosion problems in a short-sighted way. In general the initial approach is to a construction or engineering firm with a request to build a wall, or 'stop' the erosion, in some cases even to reclaim the land lost. Despite continued professional awareness of the basic concepts of coastal processes and effects of structures for at least the last 50 years (Philpot, 1984) many consultants commissioned to undertake such work continue to build walls, place gabion baskets, or construct some other form of 'hard' defence where it is not appropriate to do so.[24] This is partially due to the brief given by the coastal actor. However, the consultant has a responsibility to explain the potential effects of such structures as well as to consider whether an Environmental Assessment is likely to be required.[25] Since the recent changes, the operating authority *must* consider this in every case. The problems of such *ad hoc* decision-making in the coast can be seen in many examples (Carter, 1988: 443-458; French, 1997:75; Komar 1998: 522-534)

At present the coastal manager has four options: defend the line (hard or soft protection); managed realignment[26] (removing or breaching defences on saltmarshes and other low level coasts); managed relocation (where infrastructure is moved inland, Belle Tout lighthouse being

---

'nuisance'.

[21] *Ibid*, s. 82.

[22] *Salford City Council v McNally* [1976] AC 379, but see *Betts v Penge Urban District Council* [1942] 2 KB 154 (Div. Ct.)

[23] *National Coal Board v Neath Borough Council* [1976] 2 All E.R. 478

[24] The authors acknowledge that in some places hard defences are the only option to protect vulnerable communities. However, there are often more sustainable options in sensitive locations.

[25] 'We did not think there would be an effect' cannot always be relied upon as a suitable defence if in time the structure has an impact upon the adjacent coastal actors.

[26] The terminology 'managed realignment' is preferred to 'managed retreat' because it is value neutral.

an example of this); or do nothing. As far as environmentally friendly approaches are concerned 'do nothing' is often considered the best option for the natural coastal system. However, it does tend to introduce 'waste' into the coastal zone as well as the sediment that is required to maintain the coastal system. Protecting the coast with hard defence has the largest potential for down-drift environmental impacts. A new possibility is shown by the pioneering move of Belle Tout lighthouse where relocation allows the natural coastal dynamics to continue, whilst reducing the likely dangers to buildings which are valuable for reasons of heritage.[27] Managed realignment allows natural intertidal systems to re-establish following previous industrial or agricultural use (saltmarshes etc.). Such choices should be made in the knowledge of the sensitivity of the section of coast involved, and can be undertaken by utilising a conservation sensitive management strategy (CSMS, after Gemmell *et al*, 1996; Hansom & McGlashan, 1998).

Assessment of the economic issues relating to the construction of the defence often considers only the cost of the structure and the cost of the land protected. If it were to consider the resultant natural impacts, due to the potential risks of a down-drift occupier suing for land lost or damaged, it would have to include erosion costs. The impact of reduced beach width (loss of amenity value and wave dissipation area), and ongoing maintenance costs must also be considered. The legal duties discussed above may thus result in the incorporation of the costs of otherwise unforeseen adverse impacts to the coast in standard risk-management practice. The cost implications would also be included more widely in the process of coastal planning by actors other than local authorities, who tend in most cases to be aware of their legal responsibilities, and wary of the risks arising from litigious neighbours.

The very complexity and sensitivity to circumstance of the common law test for nuisance - on all the circumstances of the case - leads to a lack of predictability and certainty about the application of the law to any particular situation. The current common law position does, however, offer a possible means of focussing the attentions of coastal actors on the environmental impacts of their actions at the coast, and can also mandate change through effective legal remedies. It does therefore have some value from an environmental perspective. It must also be considered that scientists view the coast as a time transgressive zone, which is dynamic, rather than fixed. Legal dispute resolution is biased towards property protection, and thus is still limited, as it cannot wholly be concerned with coastal evolution, often viewing the coast as static, a line on the map. The scientific processes are only examined and considered by the law in isolated stages of the legal assessment of rights and duties, due to the historical development of the common law.

The Environmental Assessment Regulations mentioned above, as recently amended, implement European legislation to require consideration of the environmental consequences of development.[28] Coast Protection works were added to the list of projects which may require Environmental Impact Assessment (EIA) in 1994. Whether this is the case is determined by whether they are considered likely to have significant effect on the environment, and was hitherto assessed on the basis of appeal to criteria of nature, size and location, combined with informal guidance on the subject. In the recent 1999 changes the procedures have been formalised and UK

---

[27] The Belle Tout relocation is reported as involving a capital outlay of *ca* £250,000, although this must be offset against savings from the emplacement and maintenance of hard defences.

[28] Town and Country Planning (Environmental Impact Assessment) (England & Wales) 1999 S.I. 1999/293; implementing Council Directive 97/11/EC amending Council Directive 85/337/EEC.

government rates coastal development as subject always to mid-level case by case assessment. EIA is not mandatory, however neither is it subject to guidance in the form of a further minimum impact threshold.

New formalised criteria for this screening process are now given in Schedule 3 to the revised regulations and include more detailed factors. These include specifically - and selected here purely by relevance to the coast - regard to the existing land use; the regenerative capacity of natural resources; the absorption capacity of the natural environment paying particular attention to wetlands and coastal zones; the transfrontier nature of the impact; and the duration and reversibility of the impact. All of these factors militate towards the requiring of Environmental Impact Assessments in coastal projects, and are more geared towards proactive scientific assessment of the consequences than a rough and ready *ad hoc* assessment by an adversarial legal system using outdated legal constructs in an arena where interfering with the forces of nature can easily have unintended impacts on the environment. This analysis, does however, depend on the bodies entrusted with coastal management having adequate funding and authority.

Under the scheme run to date, a great deal of variation in the exercise of Council discretion existed. Typically a coast protection project required EIA if it was of more than local importance, in a sensitive area or where complex environmental impacts may well result.[29] The impact of the Environmental Assessment Regulations applies principally by way of its formalisation of processes through the planning system. The approach of the Environmental Assessment regulations embodies a search for a proactive systematic process for the resolution of disputes in advance, unlike that provided by the common law case-by-case resolution through the courts.

Although there is no specific law dealing with the process of coastal zone management, the UK has a number of voluntary partnerships, codes and advice that help to move coastal management towards a more sustainable end. The legal examples highlighted here could be used to help the current market-type system of management to become more sustainable. Most academics and conservationists would advocate a more interventionist approach by the government than the current UK system of *ad hoc* decision making (albeit often utilising a shoreline management plan). The examples outlined here would allow for the affected down-drift coastal actors to take action (or threaten action where insensitive defences were proposed), thereby ensuring that the sensitivity of new defences are fully considered, as with potential mitigation (ie. beach feeding).

The role of the law as outlined above could help provide a tool for the coastal manager or coastal actor to reduce the incidence of hard defences where it is possible there are down-drift effects. It may be that the coast is an area where more rational forms of legal regulation are appropriate, due to the advances in the science of geomorphology to the point where the legal mechanisms should look forward in the same holistic way as scientists and professional coastal experts.

---

[29] Across all forms of development the number of environmental assessments since the inception of the regulatory scheme have been greater than anticipated (Coles & Tarling, 1993), although hypothecated figures for coastal protection development since 1994 are not known to the authors.

REFERENCES

Bird, E.C.F. 1996 *Beach Management*, Wiley, Chichester.

Carter, R.W.G. 1988 *Coastal Environments: an introduction to the physical, ecological and cultural systems of coastlines*, Academic Press, London.

Coles, T. and Tarling, J. 1993 Practical Experience of Environmental assessment in the UK, *Cutting edge, 1993*, Royal Institute of Chartered Surveyors conference, *As cited in*: McAllister, A. & McMaster, R. 1994 *Scottish Planning Law: An Introduction*, Butterworths, Edinburgh.

Fleming, C.A. 1992 The development of coastal engineering, *In*: Barrett, M.G. (Ed.) *Coastal Zone Planning and Management*, Thomas Telford, London, pp5-20.

Gemmell, S.L.G., Hansom, J.D. & Hoey, T.B. 1996 The geomorphology, conservation and management of the River Spey and Spey Bay SSSIs, Moray, *Scottish Natural Heritage Research, Survey and Monitoring Report*.

Hansom, J.D. 1988 *Coasts*, Cambridge University Press, Cambridge.

Hansom, J.D. & McGlashan, D.J. 1998 Impacts of Bank Protection on Loch Lomond, *Scottish Natural Heritage Research, Survey and Monitoring Report*.

HR Wallingford 1997 Coastal Cells in Scotland, *Scottish Natural Heritage Research, Survey and Monitoring Report* No 56.

Komar, P.D. 1998 *Beach Processes and Sedimentation*, Prentice-Hall, London.

Kraus, N.C. & MacDougal, W.G. 1996 The effects of sea walls on the beach: Part 1 An updated literature review, *Journal of Coastal Research*, 12: 691-701

Kraus N.C. & Pilkey, O.H. (Eds.) 1988 The Effects of Sea Walls on the Beach, *Journal of Coastal Research*, Special Issue 4.

Mathews, 1934 *Coast erosion and protection*, Charles Griffin and Co. Ltd., London.

Motyka, J.M. & Brampton, A.H. 1993 Coastal Management: Mapping of littoral Cells, HR Wallingford Report SR 328.

Philpot, K.L. 1984 Cohesive coastal processes, *Engineering Institution of Canada Annual Conference. As Cited in*: Fleming (1992).

Tait, J.F. & Griggs, G.B. 1990 Beach response to the presence of a sea wall: A comparison of field observations, *Shore and Beach*, 59(4):11-28.

Thomson, J.M. 1997 Damages for nuisance, *Scots Law Times*, (News) 177.

Weir, T. 1996 *A Casebook on Tort*, Sweet & Maxwell, London.

LEGAL CITATIONS

A.C.- *Law Reports, Appeal Cases*; A.D. & E.L.- *Law Reports, Admiralty and Ecclesiastical*; All E.R.- *All England Law Reports*; E.G.L.R.- *Estates Gazette Law Reports*; H.L.- *House of Lords Reported Cases*; K.B.- *Law Reports, King's Bench*; S.L.T.- *Scots Law Times*; W.L.R.- *Weekly Law Reports*.

Taking an Integrated Approach Towards Flood Defence Planning in Estuaries

R.A. COTTLE, J.G.L.GUTHRIE, N.PETTITT and W. ROBERTS*
Posford Duvivier Ltd, Peterborough, England
* Hydraulics Research Wallingford, Oxfordshire, England

ABSTRACT
This paper presents the approach taken in developing strategic flood defence strategies for three estuaries on the Suffolk coast, England. As with Shoreline Management Plans the fundamental factor upon which strategic development is based is an understanding of the physical processes operating within an individual estuary. However, the greater complexity of interaction between processes, estuary form, flood defences and the estuarine environment necessitates that an iterative approach that reflects both estuary function and use is taken.

Underlying the strategic analysis is a process by which the economic costs and environmental consequences of undertaking a particular option at the local level can be effectively tracked in order to assess the implications at the estuary level. This also works in the opposite direction and enables significant impact at the strategic level to be resolved through adjustment in policy at the local level in order to achieve a strategy that meets agreed objectives and aims. Such an approach provides a strategic framework that, because of its basis on physical processes, provides a focus for the long-term planning and development of other estuary activities and uses.

INTRODUCTION
In 1998 the Environment Agency, Anglian Region commissioned Posford Duvivier and HR Wallingford to produce strategic flood defence plans for the Deben, Alde-Ore and Blyth estuaries, on the Suffolk Coast (Figure 1). This work builds upon the previous studies undertaken by ABP Research and Consultancy (1996) on the geomorphology and natural processes of the estuaries. This paper documents the approach taken in the development of the strategies and in doing so considers the relevance of the Shoreline Management Planning process to strategic flood defence planning within estuaries.

THE SHORELINE MANAGEMENT APPROACH
The Environment Agency, in conjunction with other operating authorities on the coast, have over the last few years, been involved in developing shoreline management plans (SMPs). The aim of this work has been to step back from the detailed management of coastal defences in order to take a strategic view on the need for defence. A key feature of this approach is to develop an understanding of the physical processes operating along a defined section of coastline and, based on this, assess how differing defence policies may affect the future evolution and use of the coast. As well as considering the wider coastal processes at work SMPs aim to incorporate the social and environmental functions of the coast, identifying local aspirations and expectations in the process, in order to build up an integrated framework for the management of coastal defences.

Coastal Management: Integrating science, engineering and management, Thomas Telford, London, 2000

It is clear that strategic, long term planning for defence needs on the open coast has many benefits and it has been recognised that the principles applied through Shoreline Management Planning are also desirable and applicable to flood defence management within estuary systems. In addition a physical process based approach at the strategic level provides a mechanism for investigating and dealing with the management of flood defences at the dynamic interface between estuaries and the immediate coastline. The adoption of an SMP approach to flood defence planning within an estuary, therefore, aims to develop a similarly holistic view of defence management, in keeping with, and reinforcing the approach taken on the open coast.

THE SMP APPROACH APPLIED TO ESTUARIES

Applying the SMP approach to the Suffolk estuaries raised a number of issues and challenges that previously have not been at the forefront of the strategic process on the open coast.

Process Interaction

Perhaps the most significant difference encountered is that the interaction and inter-dependence of estuarine processes and the watercourse itself is far greater than on the open coast. Any change or modification to the existing 'network of interaction' between processes and estuary form, through either natural change or human activity/influence, will act as a perturbation that feeds back through the network, altering processes and form as it does so. The potential level of influence on the interaction network will vary according to the scale at which any potential change operates. For instance, sea-level rise operates at the whole estuary level (external forcing) and it's potential influence feeds through the entire estuary regime. Smaller, localised internal change such as a modification to existing flood defences will impact most at the local level but will also have an influence beyond its immediate environment. The wider impact on an estuary system will depend on the scale of the change under consideration and the nature of any linkages within the network. Whilst interaction is recognised in many SMPs the full implications of it are often overlooked. On the open coast, where 'interaction' is generally in a single (downdrift) direction, a relatively simplistic approach is usually sufficient at the strategic level. For most estuaries however, particularly those with relatively narrow channels, an appreciation of, and allowance for, the interaction network is fundamental to strategic flood defence planning.

Applicability of Management Units

Using SMP principles to define units on the open coast has proved problematical, often throwing up anomalies in response to too rigid an application of set procedures. In the case of estuaries the complexity of interaction means that division of frontages into one level of management unit does run the risk of either ignoring local detail or missing the strategic context. Because of this and the potential interaction at estuary or local level it is considered that a hierarchical structure to examining and dividing each estuary provides the most appropriate mechanism for dealing with this issue. However, such a system must recognise the truly interactive nature of some of the changes in the processes as well as the purely one way influence which may also occur.

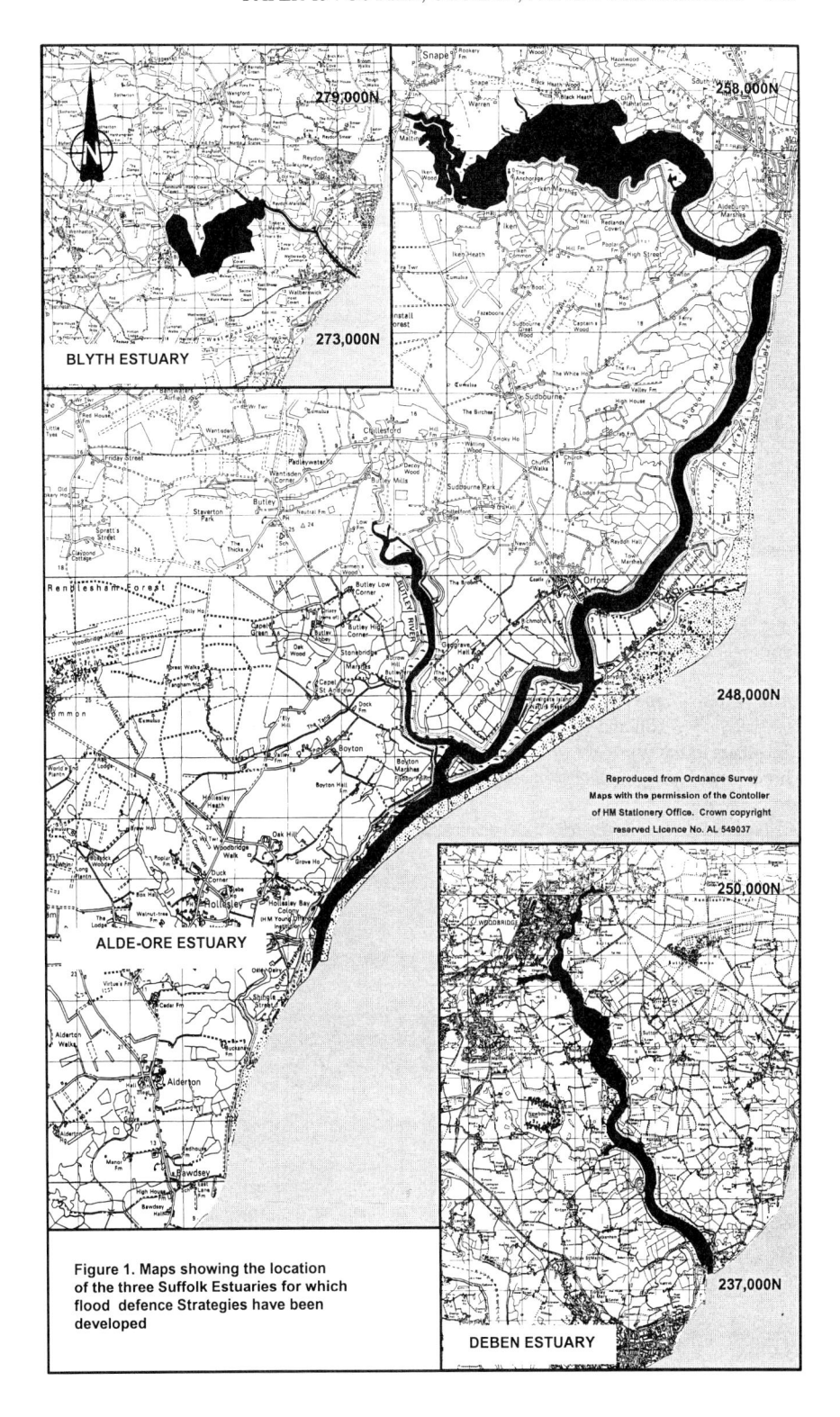

Figure 1. Maps showing the location of the three Suffolk Estuaries for which flood defence Strategies have been developed

Strategic development therefore needs to consider how every section of the estuary responds to potential change elsewhere within the system particularly as even relatively minor change could have a significant impact at the estuary level. This requires an understanding of the behaviour of each area from a physical process perspective that includes an examination of the pressures currently imposed on the area, its capacity to accommodate further physical pressure, the impact that this would have on the interests within that area and the consequences that might arise from any subsequent response. This understanding relates to two aspects:

- The physical regime (the driving forces, the response and the consequence).
- The use (the activity, economic value, defence costs and interests).

Division by Physical Regime
The practical development of strategies requires that smaller sections of the estuary are examined individually, but in such a way as to assist in building towards the larger picture. This only works if the division into smaller units is based on characteristics that reflect the interaction or linkage as a whole. The principal physical process in this respect is tidal flow and the possible response to increased flow or the control of that flow. Based on initial division on geomorphological characteristics an estuary can be divided into a number of zones, with the zones reflecting the pattern of flow through a particular reach of an estuary. In essence, this boils down to an assessment of the degree of constriction in channel width at any point within the estuary and to what degree change in the alignment of the channel at any location is constrained.

Division by use
In examining the economics of flood defence and use and interest of the estuary, and assessing how this is affected within any flood defence scenario, two main areas have to be recognised. These are the assets within the estuary channel and the assets within the potential floodplain that are currently defended against flooding. For convenience the former are generally considered on a zone by zone basis. The latter are divided by flood compartment so as to relate the cost of defence against the assets protected.

DEVELOPMENT OF THE STRATEGIC APPROACH
Based on an understanding of the physical processes the approach developed for the Suffolk estuaries effectively centres around the concept of an integrated model of each estuary against which options for individual flood compartments and strategies at the estuary level can be tested. As such the approach taken is similar to that for SMPs and can be split into two main stages. The first stage involves data collection and review and the setting of objectives. For the Suffolk estuaries objectives were derived from a number of sources including nationally accepted plans and guidelines, local strategic management plans and perhaps, most importantly, local 'aspirational' objectives obtained through consultation. The second stage effectively involves the testing of these objectives against estuary wide strategies developed via the integration of options at the local i.e. flood compartment, level. This is an iterative process that may require repeated feedback from the local to the strategic (estuary) level and back in order to achieve a strategy that is compatible with the objectives set.

Following division of an estuary to reflect physical processes and use the strategic appraisal process follows a general procedure. The implications of various options at a local level (flood compartments) are examined and consideration given as to how these may combine as options for each zone (zone options). This approach enables the impacts of options to be considered at a local level whilst ensuring that the broader implications of the various options for defence management are considered throughout an estuary. Finally, the interaction and implications of different zone options are examined to see how they work together to produce a workable strategy for the estuary as a whole.

The process is, therefore, one of predicting the future evolution of the estuary, examining how this is affected by the choice of defence options at the local level and, at an estuary level examining the consequences of this on other areas. This process is shown schematically in figure 2. Dividing the estuaries in this manner provides a hierarchical structure that enables impact to be assessed through the estuary system. Even where defences are treated in relative isolation this approach allows assessment of the cumulative impact on frontages further down stream to be undertaken. Essentially through iteration it also enables an 'impact trail' to be developed which may, where considered critical, be adjusted by amending preferred options within other critical sections of the estuary to relieve the problem.

Options

Flood compartments have been utilised as the basic building block of the strategies. For each flood compartment the cost of maintaining defences has been assessed and the damages, which would occur should defences be allowed to fail, has been determined. The current value of these damages - to property, land, and agricultural production (an important consideration in the case of the Suffolk estuaries) - has been assessed, and present values of damage derived for the future following the principles identified in MAFF's Project Appraisal Guidance Notes. Other factors such as amenity or environmental value are identified. For each flood compartment consideration is given to the generic defence policies which could be adopted. The standard strategic options considered in Shoreline Management Planning process have been utilised and are described as follows:

- "Do Nothing" - doing nothing to the existing defences and undertaking no defence work to minimise or restrict any associated damage. This option forms the basis for comparison with other options.
- "Hold the Line" - retaining the existing defence line and undertaking necessary maintenance, repairs or reconstruction as required. This option assumes that the current standard of defence is retained.
- "Managed Re-alignment" - this may take different forms. A new line of defence may be chosen, protecting key assets within the larger area of the flood compartment. Alternatively, the line of defence may be realigned or the standard of defence may be allowed to decrease.
- "Advance the Line" - this has only limited application in the estuary environment as in the majority of situations Advancing the Line would be likely to further constrict flows and increase velocities.

152 COASTAL MANAGEMENT

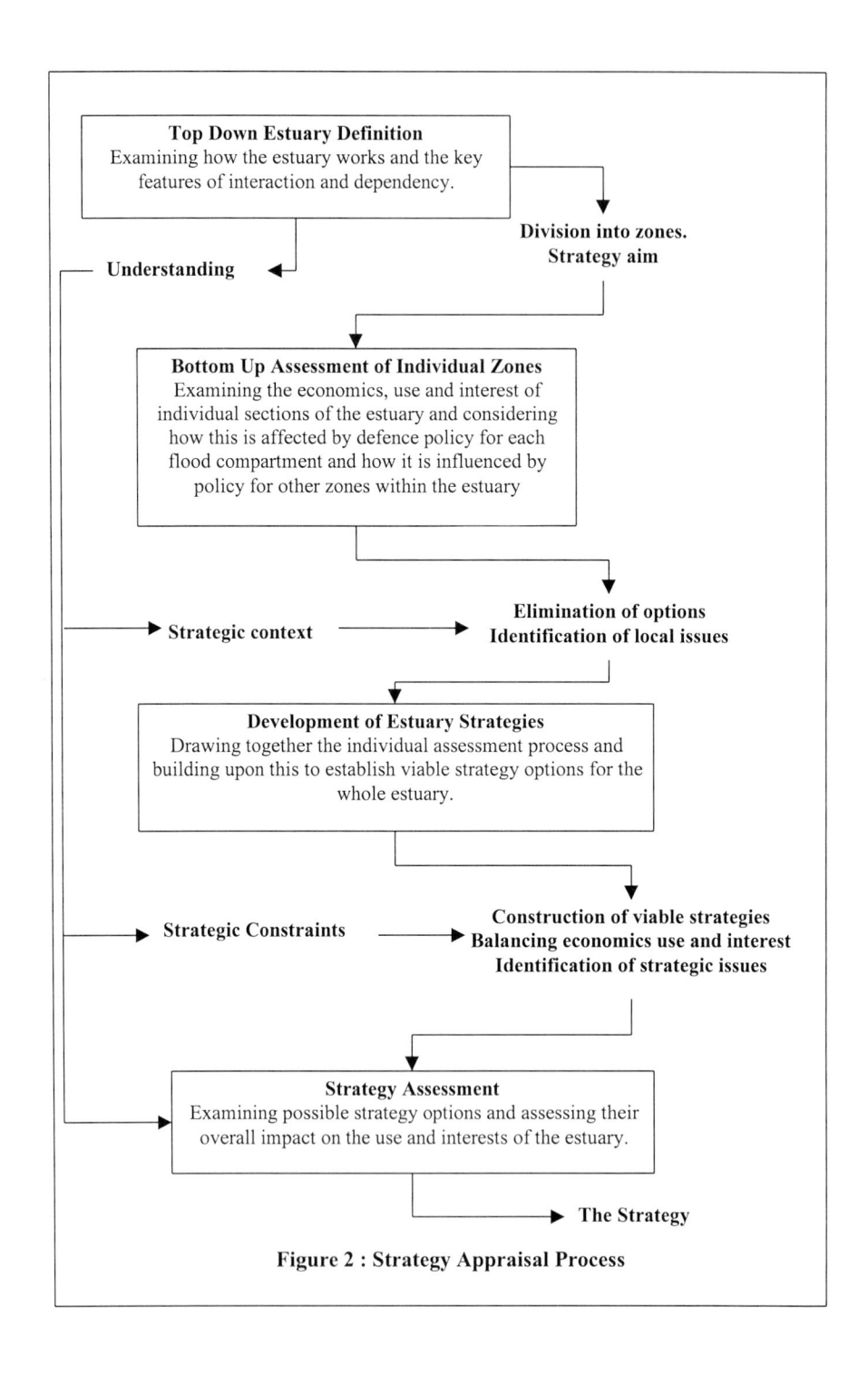

Top Down Estuary Definition
Examining how the estuary works and the key features of interaction and dependency.

Understanding

Division into zones.
Strategy aim

Bottom Up Assessment of Individual Zones
Examining the economics, use and interest of individual sections of the estuary and considering how this is affected by defence policy for each flood compartment and how it is influenced by policy for other zones within the estuary

Strategic context → Elimination of options
Identification of local issues

Development of Estuary Strategies
Drawing together the individual assessment process and building upon this to establish viable strategy options for the whole estuary.

Strategic Constraints → Construction of viable strategies
Balancing economics use and interest
Identification of strategic issues

Strategy Assessment
Examining possible strategy options and assessing their overall impact on the use and interests of the estuary.

→ The Strategy

Figure 2 : Strategy Appraisal Process

Transfer of Costs and Impacts

The decision to abandon or hold a particular defence in one area may have an impact, either adverse or beneficial, on defences or other assets elsewhere. For instance, it may appear to be economically beneficial to abandon a defence in the upper reaches of a river. However, this could lead to a significant increase in the tidal volume in the estuary, necessitating the strengthening or extension of defences downstream. Thus defence costs for the estuary as a whole could be increased. Similarly, important habitats or uses, such as in-channel boat moorings could also be lost, potentially requiring mitigation if the overall balance of use within an estuary is to be maintained. It was therefore a fundamental requirement of the appraisal process to be able to investigate the transfer of economic, environmental or social value and cost, and the potential loss or gain of interests and opportunities throughout the estuary as a whole. A mechanism for the assessment of this transfer of costs and impacts has been developed and is discussed below.

For each flood compartment (FC), there are several possible defence management options. In turn this results in several possible management options for each estuary zone as a whole; based on the different logical combinations of FC options within the zone. For any FC option, it is possible to assess the present value cost (PVc) and present value benefit (PVb) and, depending on the physical characteristics of the zone, the influence that an option may have on adjacent or opposite defences within the zone. The costs and benefits for each option may then be aggregated to provide a combined PVc and PVb for the zone under that particular combination of flood compartment options.

Under normal rules for economic appraisal the benefit cost ratio would be determined (the ratio of PVb/ PVc) and this would provide a comparative economic indicator of how worthwhile a particular option is. This method does not, however, provide any means of assessing the actual value of the option, neither in terms of its net economic advantage nor its net economic disadvantage (its deficit). If the benefit or burden of an option is to be assessed throughout the estuary, a different economic indicator has to be used, in this case net present value (NPV). This represents the difference between the PVc and PVb. If positive then the NPV demonstrates that there is an economic benefit in adopting an option; if negative there is a deficit between the cost of defending a section of the estuary compared to the value of assets protected. This indicator provides directly the value of benefit or deficit for any zone option considered. It also allows the physical impact of an option in one zone to be reflected in the economic analysis of a zone elsewhere. The NPV provides a means of tracking the economic consequence of an option throughout the estuary. Summing the NPVs for compatible options for each and every zone provides a means of assessing the economic case for the various estuary wide strategic options.

Similarly the environmental loss or gain, or the loss or gain in specific use of the estuary, may be assessed directly for any zone option and for the consequence of that zone option on other zones. In this way, and consistent with the approach adopted for the economics, a balance sheet can be maintained of loss, gain and opportunity, as the physical effect of any local option feeds through the estuary.

THE APPROACH IN PRACTICE

The strategic approach outlined above and investigation of transfer of interests and values is perhaps best illustrated by aspects of the strategy developed for the Blyth Estuary.

In its upper reaches the estuary comprises a relatively narrow channel confined within flood banks and surrounded by low lying farmland. This area is bounded by the main A12 bridge to the east. Downstream of the bridge the estuary's natural intertidal valley has been largely reinstated through abandonment of defences in the 1960s, although they still control the course of the low water estuary channel. To the eastern end of this open section the channel again becomes constrained by tidal flood defences that generally describe a series of wide bends following the original meanders of the river. The estuary channel gradually becomes further constrained by defences and runs in a virtually straight 2 km stretch before entering the sea. The entrance to the estuary is fixed by piled breakwaters. The former estuary floodplain is now largely under agricultural production with development concentrated towards the mouth of the estuary at the settlements of Southwold and Walberswick.

Within the estuary several key sections of defences are in poor condition, particularly upstream of the A12 and in the central part of the estuary. The pressure on these defences results from a combination of factors, including a lack of maintenance in the past the continuing response of the estuary regime to the large-scale failure of defences during and after the 1953 storm event and response to sea-level rise.

Based on a purely economic analysis at the local level there is a strong case for maintaining the flood defences of some flood compartments. However, the abandonment of others where it is not locally justified, would result in significant increases in tidal volume throughout the estuary potentially leading to the loss of significant environmental assets. The increase in flow would also have a substantial impact on coastal processes at the mouth of the estuary, possibly causing an acceleration in the retreat of the coastline to the south as sediment was trapped through the formation of an ebb delta at the mouth of the estuary. This option, although economically the most beneficial fails to meet the objectives and overall aim of the strategy and is therefore not considered viable.

The principal factor resulting in the unacceptability of this minimal Hold the Line option is the very high tidal flow generated by the policy of Do Nothing for the defences upstream of the A12. In isolation from economic considerations for the rest of the estuary, Holding the Line in the upstream section in order to reduce tidal flow volume, is not economically justified. However, to do so reduces the costs of defence of areas further down the estuary and in economic terms at the estuary level has the same benefit as maintaining defences where economically justified at the local level. It was found that the most effective way to implement this option is the construction of a tidal barrier at the A12 bridge. With this option there would still be significant change within the estuary, principally the loss of grazing marsh habitat designated as Special Protection Area on the southern side of the estuary and some agricultural land. However, the construction of a tidal barrier would provide the opportunity for the creation of freshwater grazing marsh upstream of the A12. As such, this option could meet compensation requirements as required under the EC Habitats Directive and enable the creation of habitat to be undertaken within an ecologically more sustainable location.

This example illustrates how both economic and environmental criteria can be transferred at the strategic level in order to achieve a strategy that meets overall aims and objectives set out at the beginning of strategy development.

In some instances the decisions taken at the local level may not influence the overall direction of a strategy. On all three estuaries there are some small flood compartments for which, based on standard economic criteria, the option would be to Do Nothing. However, due to their small size or position within the estuary, future policy has little or no impact on the estuary at the strategic level. Recommendations for the defence of these areas can therefore be made in effective isolation to the main strategy conclusions and can be based on other criteria such as local importance for wildlife or potential for intertidal habitat creation in order to replace losses that may arise within the system due to coastal squeeze. Treated as individual sites such an approach would probably not be viable under existing procedures. However, at the strategic level it is possible to make both an economic and environmental case for a flexible approach with regard to these areas and one that maintains the overall balance of interests within an estuary.

CONCLUSIONS
The complex interaction between physical processes, estuary form and use necessarily requires that a different approach to strategic flood defence planning in estuaries is taken. This paper, based on studies undertaken for three Suffolk estuaries, presents one such approach. The key to the development of the strategies has been to base strategy development on an understanding of the physical processes operating within the estuary system.

The procedures that we have adopted for the production of the Suffolk Estuarine Strategies may not be universally applicable. However, the principles of taking an integrated approach based on a process-led model can be used at all levels of strategic coastal planning and management. The development of a strategic flood defence framework enables other interests and uses, including that of habitat management at the estuary level, to also be planned in a strategic manner. As such, their implementation can be viewed as a focus for activity and provides an opportunity for change to occur. Whilst the strategies have been developed to cover a fifty year period they will be periodically reviewed to enable changes in legislation or the resolution of issues such as compensation for the loss of agricultural land, to be fully considered. The approach taken effectively enables these considerations to be built in at a later date without the need to radically alter the overall balance of function and use that the proposed defence policies support.

Shoreline Management - A Strategic Management Plan for Scarborough Town

MR. J. RIBY
Principal Engineer, Scarborough Borough Council, UK and
MR. J. HUTCHISON
Chairman of SMP Advisory Group, York, UK

INTRODUCTION

The coastline of England & Wales was at the start of the decade divided into 11 major sediment 'cells' [1]. A sediment cell is defined as a length of coastline which is relatively self-contained so far as the movement of coarse sediment is concerned and where interruption to such movement should not have a significant effect on adjacent sediment cells.

These sediment cells have formed discrete units for the development of Shoreline Management Plans [2] (SMPs). The boundaries of sub-cells are not necessarily definitive; they are based upon best available knowledge of large scale processes, and may be revised as further data becomes available. In many cases, however, sub-cells or groups of sub-cells have provided a more practical basis for the initial production of plans, being of more manageable size. Sub-cells are then further broken down into management units.

Scarborough Borough Council is the Lead Authority for the SMP for Sub-Cell 1(d) [3] which is part of 4 sub-cell plans being prepared for the Coastal Cell No.1, ie St. Abbs Head to Flamborough Head, a total length of approximately 284km. Sub-cell 1(d) covers a length of 91 km, approximately 32% of the total shoreline.

The SMP for Sub-Cell 1(d) has been operational from December 1997 and is now approaching its 2nd year in use. It continues to provide the basis for sustainable coastal defence policies along the entire North Yorkshire coastline and beyond.

In 1996 a Shoreline Management Plan Advisory Group [SMPAG] was formed to assist in the dissemination of experience gained by operating authorities from the production of the first generation SMPs. The SMPAG includes representatives from local authorities, central government, Environment Agency, English Nature and an industry representative. The development of the Sub-Cell 1(d) SMP was used as an example of Best Practice in the advisory notes issued by the group and the formation of an SMP Steering Group for Sub-Cell 1(d) continues to make use of the Best Practice advice contained within the series of advice notes [4].

One of the first priorities for the Council as set out in the adopted SMP was to prepare a long-term strategy for Scarborough town. The further development of the policies of the SMP for the coastal frontage of the urban area of Scarborough [some 7km] has involved the

Coastal Management: Integrating science, engineering and management, Thomas Telford, London, 2000

production of a coastal defence strategy for the next 60 years [5]. This strategy for the town is in fact a series of strategies covering the future management of coastal defences and coastal cliffs together with maintenance, monitoring, emergency measures [if required] and the need for further strategic studies. This has led to a prioritised programme of major capital works together with management strategies for maintenance and monitoring, totalling in excess of £39 million in the first 10 years.

The authors of this paper continue to be actively involved in the development of effective and sustainable coastal defence initiatives and this joint paper sets out the linkages between the Shoreline Management Sub-Cell 1(d) document and the Best Practice advice issued by the SMPAG. A package of strategic considerations have been adopted in order to take an holistic approach to the coast and flood protection and cliff stability issues affecting the Scarborough Town coastal frontage.

IMPLEMENTATION OF A SHORELINE MANAGEMENT PLAN

As an SMP is intended to be a 'working document' and likely to be reviewed on a 5 yearly cycle, it is important that the necessary coastal defence issues are clearly set out and prioritised in the conclusions and recommendations of the Plan. Where SMPs are in place for a length of shoreline, then MAFF will expect Flood Defence and Coast Protection Schemes submitted for grant aid to be consistent with them. This means that although the Ministry will not formally approve such Plans, they will play an increasingly important part in the approval process of schemes.

Following the production of an SMP, it is expected that individual operating authorities will develop Scheme Strategy Plans covering the Management Units within their area of responsibility. The timetable for the development of these strategies should be considered in terms of their particular priorities. For example, where there are significant coastal defence structures nearing the end of their useful lives, then it is likely that such a frontage would be one of the first priorities for the operating authority. On the other hand, if relatively new structures have been constructed indicating residual lives in excess of 10 years, then the need for a Strategy along this frontage is likely to be delayed and re-appraisal in the SMP review process in 5 years time.

In the case of Scarborough's frontage most of the defences are old [constructed pre 1900] and are reaching the end of their serviceable life or are in need of a high level of investment. The shoreline of Scarborough is throughout most of its length formed by high cliffs which still retain a legacy of potential instability which was initiated by coastal erosion prior to the construction of coastal defences during the last century. Thus the town's coastal infrastructure remains vulnerable and the Borough Council is anxious to develop and pursue a realistic, informed and prioritised strategy which will sustain and preserve the position of the town as a premier residential and seaside resort.

To assist operating authorities focus on their requirements for future strategic work, the SMPAG prepared a number of Advisory Notes, some of which have been a direct influence on the work along the Scarborough frontage. In implementing SMP recommendations, it is important to prioritise monitoring needs, capital requirements, revenue requirements and ongoing Studies and Strategies. There are a number of Advisory Notes which are intended to be of direct assistance for operating authorities in consideration of these matters, notably Advisory Notes Nos. 3, 8 and 9.

Some of the more important links between these notes and the Scarborough Town Strategy are as set out below.

1. Advisory Note 3 sets out the role of a Shoreline Management Steering Group. In this note Terms of Reference and objectives for the Group are given. Such a Group is set up primarily to oversee the implementation of the SMPs, including the implementation of any scheme Strategy Plans, such as at Scarborough. Scarborough Council, as 'lead' authority for the SMP Sub-Cell 1(d) has set up such a Steering Group, working from the paper in establishing membership and the necessary communication between this Group and the Regional Coastal Group. [A copy of the Terms of Reference for Sub-Cell 1(d) is attached as Annex 1]. It is hoped that other Coastal Groups set up for the implementation of the findings from their SMP will see the advantages and do likewise. One of the advantages of such a Group is that with wider discussions with a number of key organisations, a number of detailed survey information and research findings are coming to light, some of which were not known of or available at the time of initial SMP preparation. Obviously, using as much existing information, and working in collaboration, avoids duplication in cost and effort, as everyone on the Group are trying to achieve the common goal of appropriate coast defences.

2. Advisory Note No. 8 refers to strategic monitoring of the coastal environment, which includes the physical environment, natural environment and information on the collection, analysis and storage of coastal data. The Scarborough SMP generally sets out the necessary monitoring requirements in line with this Advisory Note. Outline of the principles for Scarborough town itself are covered later in this paper.

3. The Advisory Note No. 9 comments on the type of recommendations being made in SMPs in England and Wales and suggests an approach that could be adopted by Coastal Groups or Working Groups as specified in Advisory Note No. 3 above. It should be noted that the advice in this note focuses on the implementation of SMP recommendations and does not necessarily refer to specific individual Scheme Strategies. The paper sets out the SMP implementation requirements and suggests a methodology in taking these identified requirements further to prioritise them regionally in the form of an "Action Plan". It also suggests that one Authority is likely to be best placed to undertake or 'lead' the implementation of each action and it is hoped that workloads can be shared fairly between all organisations involved along the particular shoreline. Most importantly, it is recommended that priorities should be given to each of the actions set out in the Plan. The SMP Project Group need to ratify these recommendations in the first instance and forward the Action Plan to the Coastal Group for their area to ensure a strategic approach across Sub-Cell boundaries. For the Scarborough Town Strategy, there is no difficulty in establishing a "Lead" Authority, as there is no other operational organisation which can take on these priority studies on behalf of Scarborough Borough Council.

Other advantages of a clearly defined "Action Plan" includes the setting out of a clear programme of schemes required before the next SMP review and how they should be implemented and by which Authority, including indicative costs. Indicative costs for the next 15-20 years are also highlighted in this SMP, which allows the operating authority and MAFF to have a better indication as to the long-term financial needs for an area.

Provided there is good liaison between the SMP Project Group, with all parties knowing exactly what needs to be undertaken and when, progress in the implementation of the recommendations as set out in any SMP should be relatively straightforward and achievable. It should be noted that if the recommendations are to be undertaken and completed before the next review begins, then each organisation will need to take on their defined role seriously and diligently.

Currently, the Project Group for Sub-Cell 1(d) meets twice a year to discuss progress and long-term planning generally. It is the intention of this group to prepare an annual report which will set out the actions in the SMP and what successes have been achieved in the interim. It also provides an indication as to costs likely to be required in the next 3 years, which also assists in MAFF's Forward Planning and focuses the necessary planning for each operating authority to ensure that it has submitted the necessary long-term plans to MAFF, eg LDW 11 forms.

SCARBOROUGH TOWN STRATEGY

Plate 1 : Aerial View of Scarborough Coastal Frontage

The coastal defences of the Scarborough urban area in general terms consists of a harbour, a series of sea walls and breakwaters whose main purpose is to protect the development within the coastal margin from the forces of marine storm damage and flooding. Most of the defences are old and are either reaching the end of their serviceable life or are in need of a high level of regular maintenance. The shoreline of Scarborough is throughout most of its length formed by high cliffs which still retain a legacy of potential instability which was initiated by coastal erosion prior to the construction of those defences during the 19th century and the early part of the 20th century (see Plate 1). Therefore the town's coastal infrastructure remains vulnerable to an ageing coastal defence system backed by cliff slopes which have the potential to be unstable to varying degrees.

A coastal defence strategy study was undertaken by a team of experts in coastal studies from the Council's consultants, High Point Rendel, supported by the HR Wallingford. The study involved a review of available data compiled within the SMP and other relevant sources. It has sought to collect additional data to assist in the development of the strategy. In these terms it has reviewed the condition, performance and residual life of all the existing coastal defences in some detail. It has produced a 'model' for the analysis of overtopping performances of the existing sea walls and breakwaters and it has involved extensive site investigation of particular sites in the South and North Bays, including the installation of monitoring equipment. Analysis and interpretation of the information to define the nature of the problems along each part of the coast and to identify appropriate management strategies was carried out and this has involved the following:

i. a quantitative risk assessment of sea wall and breakwater failure as a result of wave impact forces and structural conditions;

ii. an assessment of the consequences of sea wall and breakwater failure;

iii. an assessment of wave overtopping problems;

iv. a quantitative risk assessment of the threat to the sea walls from landslides;

v. preliminary environmental assessment of the defence options; and

vi. an identification of the coastal defence options for the next 60 years.

A number of fundamental principles have been established as part of these strategic studies to provide a framework for the development of the management strategy. These are:

a. to undertake urgent improvements to the condition and performance of high risk sea walls and breakwaters;

b. to make allowances for sea level rises in the design of improved sea walls and breakwaters;

c. to undertake a regular programme of sea wall and breakwater inspection and maintenance;

d. to make contingencies plans for prompt emergency response to rare but potentially very serious problems, for example sea wall breakages;

e. to adopt a precautionary approach in tackling the problem of residual landslide risk involving regular monitoring, slope inspection and maintenance;

f. to develop contingency plans for emergency stabilisation measures should the slope monitoring and inspections reveal evidence of a deterioration in the stability; and

g. to ensure that all elements of the management strategy enhance or maintain the quality of the coastal environment and take account of the broader coastal management objectives.

The length of coast between Holbeck and Scalby Mills was therefore divided into 15 individual sections (see Figure 1) and the proposed action for these sections involving a combination of specific strategies related to sea walls, structural performance, wave topping and cliff instability have been produced. Arising from these studies are a series of significant capital works prioritised in order of risk against failure.

The three most urgent areas at Scarborough town are the East Pier of Scarborough Harbour, the Marine Drive including the Holmes and the Spa Complex including the Spa Chalet cliff. These three sites are in high risk/exposure locations and often take the brunt of the north-easterly seas that drive against Scarborough's frontage. This has been reflected in the number of emergency coast protection schemes carried out on these structures in the last 5 years.

As well as these particular sites the strategy identifies others along the coastline in a series of prioritised measures. The strategy has highlighted that significant capital investment will be required in the next 5 to 10 years in order to sustain the existing line of sea defences and cliffs. As well as capital works the strategy has also identified monitoring regimes and foreshore and cliff management strategies as well as a maintenance strategy. Most of these capital improvement works and long-term monitoring requirements should be eligible for grant aid from MAFF.

ADVANTAGES OF A STRATEGIC APPROACH

The preparation of detailed Scheme Strategies for locations such as Scarborough has been a long-term objective of MAFF. A similar strategic approach could be applied to many sites in England. However, in order to assist operating authorities to prepare such strategies, MAFF provided Interim Guidance on the preparation of strategies in 1997 [6]. This document sets out clearly the intentions of a Scheme Strategy and, again, the Scarborough Town Strategy can be considered as a good example for the following reasons.

1. The problems at Scarborough are set out in an integrated way which considers solutions within a time frame. (The problems and the solutions are within a relatively self-contained area).

2. It provides a means for establishing and justifying the overall aims and objectives of flood and coastal defence policy for the area concerned.

162 COASTAL MANAGEMENT

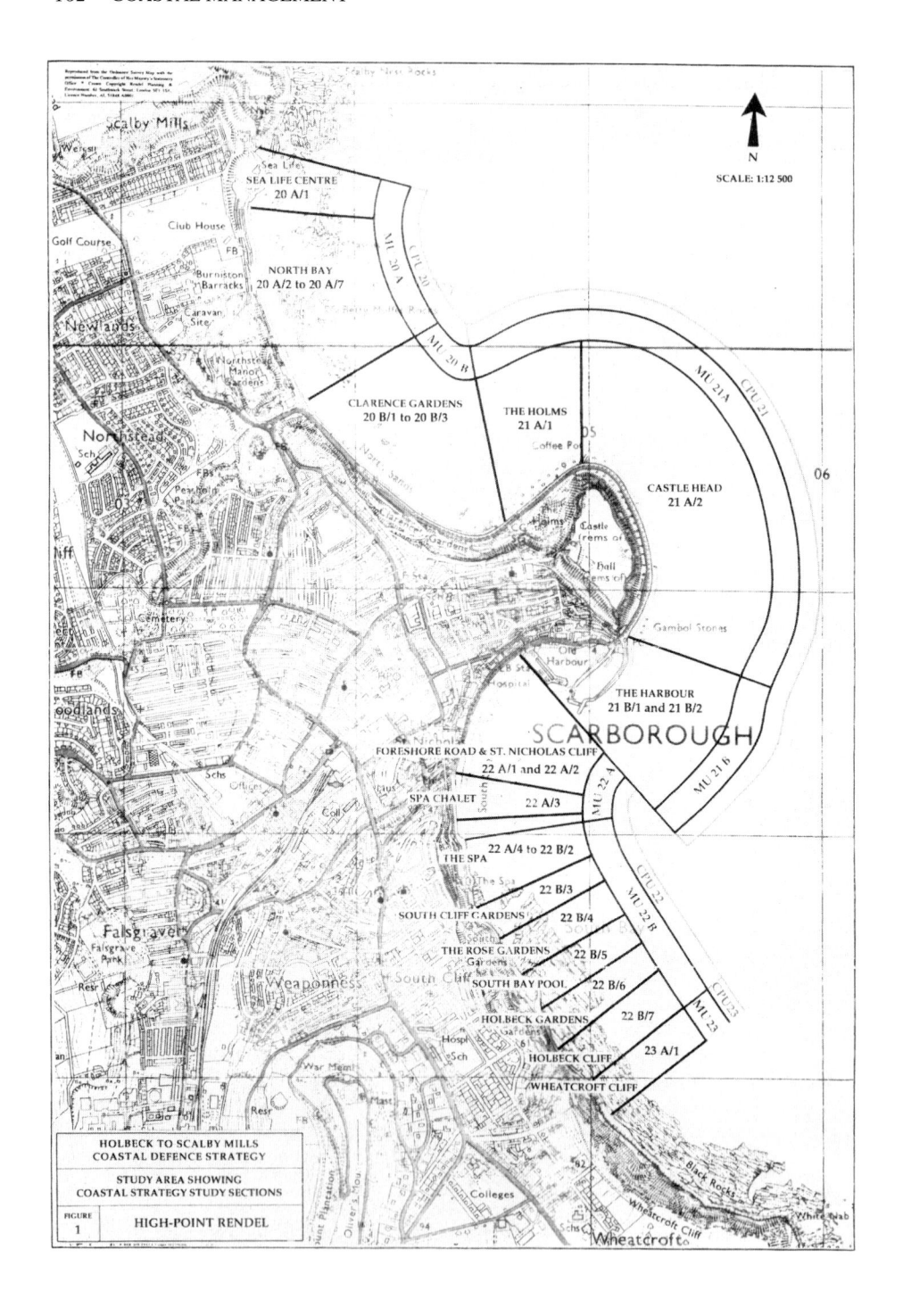

Figure 1. HOLBECK TO SCALBY MILLS COASTAL DEFENCE STRATEGY — STUDY AREA SHOWING COASTAL STRATEGY STUDY SECTIONS. HIGH-POINT RENDEL.

3. It attempts to concentrate on the generic principles for the achievement of these aims and objectives, and sets out the logic for transparent decision making.

4. The scale of the problem is clearly identified and the Strategy sets out a methodology for the delivery of specific engineering schemes.

5. It sets out in detail the economic, environmental and technical assessments at a strategic level which take account of all the relevant impacts and opportunities which can later in the approval process be applied at the scheme level.

6. It provides for effective feedback so that lessons from scheme implementation which have an impact at a strategic level in development in knowledge and understanding can be incorporated into subsequent reviews of the overall strategy (and, hence, the SMP itself).

7. A clear programme of long-term works are set out which allow the operating authority and MAFF to forward plan for the identified sums of money together with the prioritisation scoring for the schemes listed.

8. It explores opportunities for cost effective methods of procurement of works. It should be noted that such opportunities as linking a number of coastal defence schemes on the coast can be considered, including linkages with development opportunities within the coastal margin which would underpin the coast defence element of the schemes.

9. It allows for a clear and definable consultation process so that all stakeholders are aware of the long-term impacts of the scheme or proposed schemes.

10. It allows MAFF agreements to be obtained for proposals at Scarborough which should allow for simpler and more straightforward submissions in the longer term.

CONCLUSIONS

The Scarborough Town Strategy has benefited greatly from the guidance provided by the SMPAG. It has enabled a coherent strategy to be developed dealing with each of the shoreline elements in an integrated way.

The proposed schemes, prioritised in an informed manner having regard to level of service and risk of failure provide the Borough Council and MAFF with a structured approach to delivering the coast protection service for Scarborough town.

The strategy recognises the scale of problems to be addressed and underpins major capital works with strategies aligned to monitoring, maintenance and ongoing investigations and further studies.

The priority actions for the implementation of the Scarborough Town Strategy will now involve:

1. Detailed engineering solutions for each of the prioritised capital schemes, including environmental assessment and economic evaluations.

2. The further development of monitoring, inspection and recording procedures for coastal cliffs and sea walls.

3. The development of a structured and prioritised maintenance programme based upon critical sections.

4. The establishment of contingency plans for emergency slope and sea wall works, pending major capital works.

5. The undertaking of further detailed investigations and studies to assist the development of sustainable engineering solutions on vulnerable sections of the coast.

6. A review of the available methods for the procurement of capital works (including investigation of the use of central government's Capital Modernisation Fund) [7].

7. A 5-yearly review of the management strategy, taking into account changing circumstances and priorities and improved understanding of the coastal and slope processes.

7. The strategy will inform the evolving Shoreline Management Plan.

It is the intention of Scarborough Borough Council to apply this approach to other locations along the North Yorkshire coastline and similar commissions are planned for the other centres of human settlements including Whitby and Filey as well as the undeveloped coastline where environmental management, including the natural coastal defence of foreshore outcrops and beaches are prevalent.

ACKNOWLEDGEMENTS
Both authors would like to thank Dr. Alan Clark and Mr. Steve Guest [Scarborough Town Strategy] for their assistance in the development and understanding of the coastal defence issues at Scarborough town. It should be noted that the views in this paper are those of the authors and do not necessarily reflect any particular organisation.

REFERENCES
1. Coastal Management: Mapping of Littoral Cells. Report SR 328 Hydraulic Research, Wallingford. January 1993.

2. Shoreline Management Plans: A Guide for Coastal Defence Authorities. MAFF (1995).

3. Huntcliffe (Saltburn) to Flamborough Head Shoreline Management Plan prepared by Mouchel Consulting Limited; September 1997.

4. Shoreline Management Plans: Interim Guidance prepared by the Shoreline Management Plans Advisory Group [Advisory Notes Numbers 1-12]. MAFF 1998.

5. Holbeck - Scalby Mills, Scarborough - Coastal Defence Strategy prepared by High-Point Rendel.

6. Interim Guidance for the Strategic Planning and Appraisal of Flood and Coastal Defence Schemes. MAFF [1997].

7. H.M. Treasury Public Expenditure Systems PE (98) 37 - Capital Modernisation Fund.

ANNEX 1

THE ROLE OF A SHORELINE MANAGEMENT PLAN

STEERING GROUP FOR LITTORAL SUB CELL 1D

1. **Terms of Reference**

   - To oversee the implementation of the SMP.
   - To develop and oversee implementation of Scheme Strategy Plans contained within the SMP.
   - To monitor progress of all initiatives proposed within the SMP.
   - To initiate formal reviews of the SMP.
   - To liaise closely with the work of the North East Coastal Authority Group (NECAG) on all shoreline management issues.
   - To liase with MAFF Regional Engineer on future proposals.

2. **Objectives**

   The objectives of the SMP Steering Group are as follows:

   - Maintain an appropriate size (expertise and number) of SMP Steering Group giving full consideration to the natural processes at work in the area (and appropriate boundaries) for the future management of coastal defences.
   - Assess the SMP linkages with other Plans (e.g. Heritage Coast Strategies, LEAPs, Statutory Plans, etc) to ensure that conflicts are minimised and opportunities maximised.
   - Establish a framework to prepare detailed scheme strategies, a framework for action and for future SMP reviews. (Some strategies may involve applications under both Flood Defence and Coast Protection legislation. This should not cause difficulty, but it is advisable for one 'lead' authority to obtain overall MAFF agreement to the strategy.
   - Co-ordinate and develop mechanisms to implement and deliver the adopted SMP policies.
   - Steer and review the outputs of the SMP so that its agreed objectives are met.
   - Ensure consistency on all aspects of shoreline management planning.
   - Identify the appropriate scale of the problem and provide the necessary guidance for delivery of specific engineering schemes to the operating authorities.
   - Review the conclusions of the SMP and prepare briefs for the next review.
   - To consider the strategic aims and objectives for implementation which should be established jointly with all key stakeholders.
   - To promote liaison and co-operation between members of the SMP Steering Group, Heritage Coast Officer Working Groups and other interested organisations and parties.
   - To assess the management units which pose the greatest problems and review the Steering Group representation as necessary to reflect key user interests within those management units.

- To assess how scheme strategy plans can be promoted where management units cross administrative boundaries.
- Any additional studies, information, monitoring or local research required to fill gaps in current knowledge should be established, together with identification of responsibility and a 'lead' authority to do this work. Funding sources should be identified.
- To aid the decision making process so that they key issues and potential conflicts of interest are properly handled.
- To stimulate interest in the future management of the coastal defences in the area.
- To develop the consultation process which remains an essential element of scheme strategy development and requires careful preparation and management.

3. Membership of Group

- Chairman, Mr. J. Riby, Scarborough Borough Council.
- Technical Secretary, Mr. N. Corrie, Scarborough Borough Council.
- Mr. J. Hutchison, Regional Engineer MAFF (York) - (Observer).
- Mr. C. Matthews, East Riding of Yorkshire Council (Beverley).
- Mr. G. Stamper, Redcar and Cleveland Borough Council.
- Mr. R. Martin, Environment Agency (York).
- Mr. M. Welbourn, North York Moors National Park (Helmsley).
- Mr. S. Pasley, Countryside Commission (Leeds).
- Mr. D. Clayden, English Nature (York).
- Mr. D. Williams, Planning Division, Scarborough Borough Council.
- Other consultees or stakeholders should be invited to group meetings when appropriate to discuss their issues of concern.
- MAFF's Regional Engineer will be kept informed of all group meetings and invited to attend as an observer.

4. Reporting and Communication

- The numbers of meetings will reflect the needs and the key issues within the SMP boundary and the numbers of strategies being developed.
- NECAG will receive a copy of the minutes of the sub group meetings, together with brief reports on key implementation issues.
- The Chairman of the group should provide a written update at NECAG meetings on all progress and issues.

The Bristol Channel Marine Aggregates Resources and Constraints Research Project: Keeping the Consultation Process Clear of Murky Waters!

JAN BROOKE, Environmental Consultant,
Market Deeping, Peterborough, UK

INTRODUCTION
The Bristol Channel Marine Aggregates (BCMA) Resources and Constraints Research Project is a major piece of research being sponsored by the Welsh Office, DETR and the Crown Estate. These organisations require strategic level information in order to assist them with the assessment of dredging licence applications, and with determining the possible wider implications of dredging activity. The BCMA project is therefore concerned with, *inter alia*, coastal processes, physical characteristics and environmental interests throughout the Bristol Channel. For the purposes of the project, the study area covers the length of the Bristol Channel defined by lines drawn across the River Severn at Newnham, Gloucestershire and across the Channel from St. Ann's Head (Dyfed) to Hartland Point (Devon).

The overall aims of the project are:
- to further develop our understanding of the sediment transport regime in the Bristol Channel and the extent to which the sediment deposits are interlinked;
- to define the marine aggregates resources and to evaluate constraints on their extraction in the Bristol Channel; and
- to prepare a report on the above to assist those organisations involved in the Government View procedure (and any subsequent arrangements) in the evaluation of proposals for future dredging.

BACKGROUND
When the BCMA project commenced in 1996, members of the study team* were aware that much of the existing information available related only to certain aspects of the Bristol Channel (for example, information collected as part of site specific coastal defence schemes, dredging licence applications, etc.). Very little information dealt with physical and environmental interests at a strategic level. Some such data were being collated for the preparation of shoreline management plans, but as several such plans fall within the boundaries of the BCMA study area, even this process was not culminating in all relevant information being held in one place. An objective of the BCMA project was therefore to bring together all relevant existing data, to identify critical data gaps and (within the financial constraints of the study) to collect new data. Without such strategic level information and the resulting improved

Coastal Management: Integrating science, engineering and management, Thomas Telford, London, 2000

understanding of how the natural system works, it is difficult to make decisions on whether a specific area of sand or gravel can be dredged without causing harm to the environment.

Dredging in the Bristol Channel currently supplies, *inter alia*, 95% of the materials required by the South Wales construction industry. Demand for materials for construction and beach nourishment in both South Wales and South West England is high and is predicted to remain high into the foreseeable future. There is concern amongst some groups, however, that dredging activity may be causing increased erosion at the coast and environmental degradation in both marine and coastal areas. The procedure by which previous dredging licence applications have been assessed has also been criticised for its lack of transparency. Overall, the issue of dredging in the Bristol Channel is therefore very sensitive.

THE CHALLENGE
There are more than 300 organisations and groups with an interest in aspects of dredging activity in the Bristol Channel. These range from the Environment Agency, the English and Welsh countryside agencies and other organisations represented on the Steering Group, to parish councils and community groups. Many of these organisations and groups have strong views about dredging, particularly the possible local effects of dredging. The project Terms of Reference, however, made clear that the research was not intended to deal with local or site-specific issues. Its aim was to significantly improve the strategic level information and understanding available to those responsible for making decisions on dredging. With such a laudable objective, it would be very disappointing if any misunderstandings about the scope and purpose of the project remained at the time of publication of the final report.

Communicating with such a large number of consultees was therefore an important and potentially quite daunting task, not least because of the need to ensure that the strategic nature of the research project was understood. It was essential to ensure that unrealistic expectations in respect of the project outputs were avoided, right from the outset of the project.

Against this background, the project team considered effective communication to be a high priority element of the project. An open and transparent approach needed to be adopted, designed to generate and maintain confidence in the team's approach. Interested parties needed to be able to follow the progress of the project, and also to ask questions and feed in information and ideas at each stage. However, the nature and scale of the consultation process had to be balanced against the cost: all parties agreed that the bulk of the available funding had to be spent on the research itself. The challenge facing the team was therefore to implement an effective, wide scale consultation exercise without significantly diminishing the funds allocated for the research. Detailed planning and the use of carefully selected techniques were of vital importance.

INITIAL CONSULTATION PROCESS
The first stage consultation exercise involved the publication and circulation of a six-page Consultation Document which included relevant maps/figures of the study area and provided contact details for the study team. The Consultation Document

introduced the project and its objectives, set out the study team's proposed approach, and highlighted the intended study outputs. Within this context, the document invited respondents to advise the study team of any available data which might be of interest to the project and also of any issues which they felt the project could usefully address. The document provided respondents with the opportunity to request a meeting with a member(s) of the study team, if they felt that such a meeting would be worthwhile.

The document was circulated to all groups and organisations known to have an interest in the study area. This included not only the statutory agencies but also other organisations and groups who had previously been involved in correspondence (for example with the Welsh Office) about dredging and its potential effects. All of those contacted were asked to raise awareness of the project amongst other potentially interested parties and, as a result, additional requests for copies of the document and/or inclusion on the project mailing list were received. A consultation database was set up and managed using MS Access software: all consultees' details were maintained using a "tracking database". This ensured not only that details could be updated as necessary, but also that the study team could organise the database according to the particular interest(s) of the consultee organisation.

The circulation of the initial Consultation Document was followed by a series of meetings. In some cases, these meetings were requested by consultees; in others, they were initiated by team members. In all cases, however, a common procedure was followed in order to ensure that the meeting was as productive as possible.

The study team members attending each meeting were selected to ensure that their disciplines matched the interests of the consultees (eg. specialists in nature conservation or coastal processes). Following an introductory presentation at the meeting by the team member(s), there were then two key requirements. The first was to identify available information/data which would potentially be of use to the study: the team needed to know the type of information and how/by whom it was held. The second requirement of the meeting was to ensure that consultees were able to raise issues which they felt that the study should address. When local rather than strategic issues were raised by consultees, this gave team members the opportunity to reinforce the strategic objectives of the study, to clarify the scope of work which would be covered (as defined by the Terms of Reference) and to explain why such local issues were not part of the brief.

The majority of meetings involved representatives from a number of groups with a common interest(s). In some cases these were geographic, in others they were discipline or topic-related (eg. Regional Aggregates Working Parties). One-to-one meetings do not generally provide such a good opportunity to optimise use of the available budget. Further, it was found that meetings involving representatives of a number of groups were more productive as a wider variety of issues were generally raised and discussed amongst the interested parties, thus reducing the need for follow-up correspondence/meetings.

NEWSLETTERS
Once the project was underway, one mechanism used to keep interested parties up-to-date with progress with the project was the preparation and issue of project

newsletters at approximately six month intervals. Each newsletter included the following information, accompanied by a specific request for feedback:
- progress since the previous newsletter
- key findings
- ongoing and planned work
- proposed future consultation, including details about forthcoming consultative seminars (see below).

The first newsletter, for example, listed all the issues raised during the initial round of consultation, together with a brief explanation of the extent to which each issue would be dealt with under the Terms of Reference for the project. Readers were therefore able to recognise issues or questions they had raised and were aware of whether or not the project output would deal with their concerns. Subsequent newsletters included, *inter alia*, summaries of the findings of the project's Interim Report and of the main results of the primary research (ie. original data collection) undertaken during the second phase of the project. Where potential information gaps were identified, requests for additional information were also made via the newsletters. However, it is worth noting that discussions at meetings and at the consultative seminars (see below) were generally more successful in terms of identifying and gaining access to the available information than were requests via the newsletters.

At the time of writing, one newsletter remains to be prepared - the project's concluding newsletter. This newsletter will present the findings of the final report, describe the methodology developed to assist in the decision-making process, and provide details of who to contact in order to find out about progress with the implementation of the project's recommendations.

CONSULTATIVE SEMINARS
In addition to the issue of newsletters, a series of consultative seminars was also held. Each seminar comprised an introduction and short presentations by members of the study team (dealing, for example, with data collection, modelling, resource evaluation, and environmental interests) followed by question and answer sessions. The verbal presentations were supplemented by visual information including graphics prepared and presented using the PowerPoint software. The seminars were held at large conference hall venues where the available facilities included cinema style screens for clear visual projection and good quality acoustics. The seminars were open, by invitation, to representatives of all the groups consulted.

The objectives of the consultative seminars were as follows:
- to enable the study team to present the results of progress with the project to all interested parties, without incurring the costs associated with small group meetings (whilst these had played an important role during the initial stages of the project, they were not felt to be justified at later stages)
- to provide consultees with direct access to all members of the study team and hence to all disciplines (the seminar approach thus provided interested parties with an advantage because it would not usually be cost-effective for more than two team members to attend a smaller meeting)

- to provide an opportunity for interested parties to ask questions of members of the study team, either from the floor during the seminar, or informally, following the seminar.

Whilst careful forward planning and preparation were necessary, this approach generally proved very successful. In each case, the venue for the seminar was provided by one of the interested groups. The consultants team, meanwhile, provided all the presentation materials. All consultees were invited to attend the seminars (although, interestingly, several chose not to attend on the basis that they felt sufficiently well informed about progress with the project as a result of receiving the newsletters). The seminars provided the study team with an opportunity to reinforce the aims and objectives of the project and to present the results to-date in that context. The open-forum approach helped to ensure that team members were aware of the issues and concerns associated with dredging in the area, even when these concerns were not being tackled by the study. Finally, the seminars also proved to be good testing and motivational events for the study team.

Each consultative seminar took place during a full afternoon session with team members available as necessary for follow up conversations on a one-to-one basis after the formalities of the seminar had been concluded. Additional evening meetings following the same format were also offered (for example, to cater for those representing NGO (non-governmental organisation) interests and community groups). Although the vast majority of consultees in fact opted to attend the afternoon sessions, it was important to ensure that none of the interested groups felt that they were being excluded.

At the time of preparing this paper, the two final dissemination seminars, one in South Wales and one in south-west England have yet to be held. The purpose of these seminars, to which all interested parties will again be invited, will be to present the results of the research project. As with the previous seminars, an interactive approach will be adopted. The study teams' presentations will be followed by question and answer sessions, thus enabling any additional explanations or clarifications to be provided.

MEDIA INTEREST
Finally, it is worth mentioning that members of the press, particularly the local press, showed an interest in the project. The Bristol Channel Marine Aggregates project featured on John Craven's Countryfile and local BBC radio, and in a number of specialist magazines. There were also various articles in local and regional newspapers. In addition to providing information to the media, the project team were involved in monitoring coverage in the local press and, where necessary, responding (eg. to published "letters to the editor") in order to ensure that there was no misunderstanding about the background to the project and its objectives.

UNDERLYING PRINCIPLES AND LESSONS LEARNED
Various important principles, based on a "communication ethic", have underlain the approach adopted by the BCMA study team.

Openness and transparency were of fundamental importance in communicating what the project was trying to achieve, its objectives and the methods and techniques being used, and also in generating confidence in the study team.

Careful planning and programming of the consultation process, including the identification of which techniques to use at which stage in the study process, was essential. The process also needed to be carefully managed - for example, the project manager provided a single point of contact for both consultees and members of the team alike in order to avoid confusion and duplication.

Just as important as these, however, have been the principles of involvement and integration - both internally and externally.

Internally (ie. within the team), every effort has been made to involve representatives from all the project disciplines in the consultation process. In addition to the standard practices of discussing issues raised by consultees at team meetings and copying correspondence to other members of the study team, representatives from all the key project disciplines became directly involved with interested parties. They attended the initial consultation meetings and the consultative seminars. They also provided input into the newsletters. Awareness of the issues of concern to consultees was high amongst all members of the team.

Externally, members of the study team have tried hard to ensure that interested parties feel that this is, at least in part, their project - that they are able to contribute and to get involved. At each stage, information has been provided to enable consultees to develop an understanding of the project's objectives and of the investigative methods and techniques being used. The team has been accessible and has tried to respond promptly when questions have been asked and issues raised. Throughout the project, correspondence has been encouraged.

The BCMA project still has a few months to run but the approach adopted to the consultation process has, thus far, been well received. Marine aggregates dredging is currently a "hot topic" in terms of coastal management, particularly in the Bristol Channel area. The approach adopted, and the lessons learned, by the BCMA project team are likely to be both interesting and relevant to others responsible for managing the consultation component of all forms of large scale coastal management project.

ACKNOWLEDGEMENTS

The author wishes to thank the following for their advice and assistance in the preparation of this paper: Charles Haine, Posford Duvivier Environment; Bill Cooper, ABP Research; and Chris Morgan, Welsh Office. The views expressed in this paper nonetheless remain those of the author and do not necessarily represent those of the other organisations involved in the project.

* The study team for the BCMA project comprises Posford Duvivier and ABP Research, supported by various specialists. The author, formerly Manager of Posford Duvivier Environment but now a freelance consultant, is continuing to work as a specialist advisor to the project on matters relating to consultation.

The Changing Flood Risk Requirements and the Managed Retreat Policy – Dungeness Nuclear Power Station and how it might be Applied Elsewhere.

DR ROGER MADDRELL and BILL OSMOND
Halcrow Group Limited   Nuclear Electric
Swindon, UK   Dungeness, UK

ABSTRACT

The two reactor buildings of Dungeness Nuclear Power Station (the Station) were built in the late 1950's on the southern, eroding coast of Dungeness, which is the largest coastal shingle feature in Europe, having an area of some 35km^2. The site is exposed and was chosen because of its proximity to deep fast flowing water, but local erosion of the station's frontage was up to 1.5m annually. Despite the increasingly stringent flood risk requirements, the Station has been protected from erosion throughout its life by beach feeding. The site is therefore a good long term example of a 'soft' defence protecting a high risk area.

The shingle for the beach feeding has been taken from the accreting downdrift eastern shore annually during winter months. The suitability of this method of maintaining coast stability and the flood protection it offered, has been regularly reviewed and has always been found to be considerably less expensive than the more formal and 'harder' options. However, the amount of shingle required to maintain the beach was increasing over the years as the main feed point became more and more out of regime with the updrift coast. Consequently, a new policy was adopted that allowed the coast to retreat locally within the site. This has over a period of five years produced considerably cost savings, while still allowing the level of flood protection to be increased and reducing the impacts of shingle extraction downdrift. It has also improved beach stability in critical areas.

The most recent safety review examined the risks and requirements associated with extreme events, up to the 1 in 10,000 year tsunami. As a result, the main shingle flood embankment level had to be increased to + 8m OD and its crest width to 20m and modifications had to be made within the Station and to its gates. While these additional shingle quantities were taken from the existing borrow area, the quantities involved were less than those required prior to the implementation of the managed retreat policy. Consequently, coastal changes downdrift of the borrow area were able to be limited.

Coastal Management: Integrating science, engineering and management, Thomas Telford, London, 2000

The paper discusses the consequences of adopting a coastal managed retreat policy in an exposed coastal area where the risks are perceived to be high, the longer term impacts of the original "hold the line", the new managed retreat policy and how this policy might be applied to other sites in the UK with varying types of coast.

1 DUNGENESS

1.1 Introduction

Dungeness, on the SE coast of England, was formed initially as a shingle bar across Romney Bay with rising sea levels at the end of the Pleistocene, some 5300 to 6000 years before present (BP). Estuarine deposits accumulated landwards, now forming the marshes, while shingle continued to accumulate on the seaward side to form the Ness (Greensmith and Gutmanis, 1990), Figure 1 shows the postulated coastlines from the Neolithic (11,000 to 4,000 years BP), from when the upland areas between Rye and Hythe was the cliff line, to the present (after Lewis, 1932).

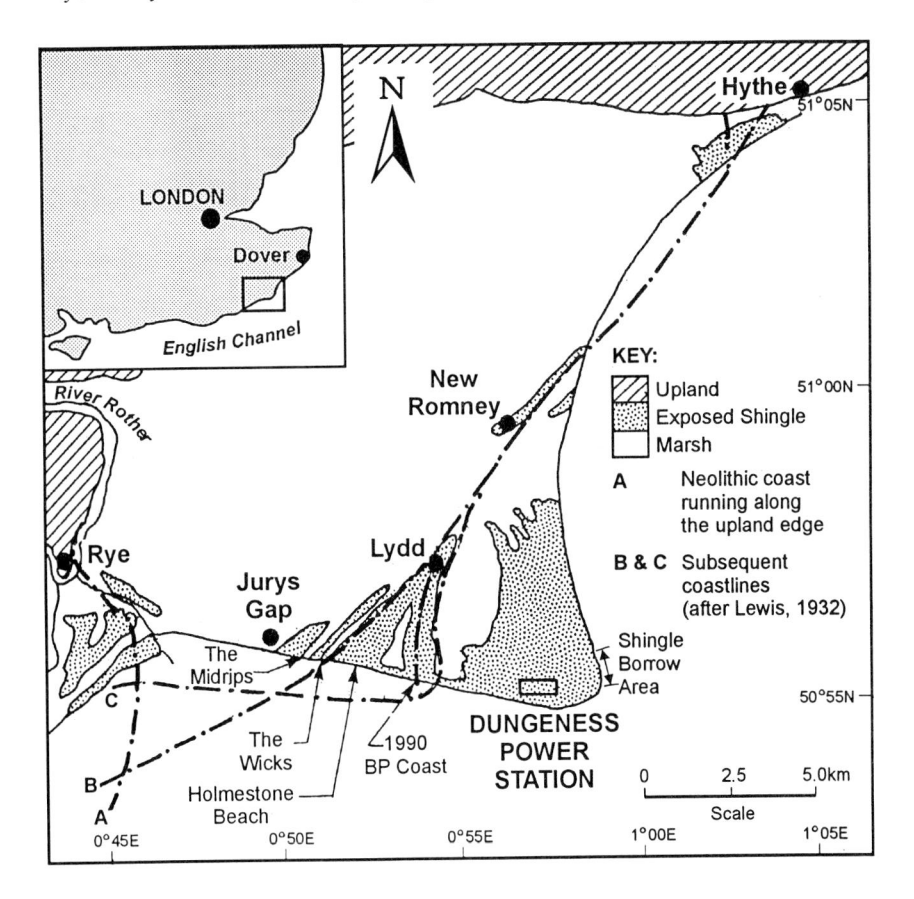

Figure 1 – Site Area

The past and present alongshore drift is as a consequence of the dominant southwest waves moving shingle eastwards, eroding the southern coast and causing accretion along the eastern side (see Figure 2). The more recent morphology of the Ness, first described by Lewis (1932), is clearly evident from the old storm shingle ridges, whose size and height are thought in part to reflect the changing sea levels, climate, wave energy and the supply of shingle.

Sea-level was relatively stable between 4000 and 2000 years BP, but during the last 2000 years it rose by about 2.5m as evidenced by the height of the ridges. On the south coast, the oldest ridge levels at Jurys Gap are the lowest, generally rising some 0.2m to Homestone beach and remaining at a similar level (4.3m OD) as far east as the 1900 BP coastline. The present storm ridge is about 1.5m higher in this area, equivalent to the 1300 BP coast levels, when the climate was more severe (Lamb,

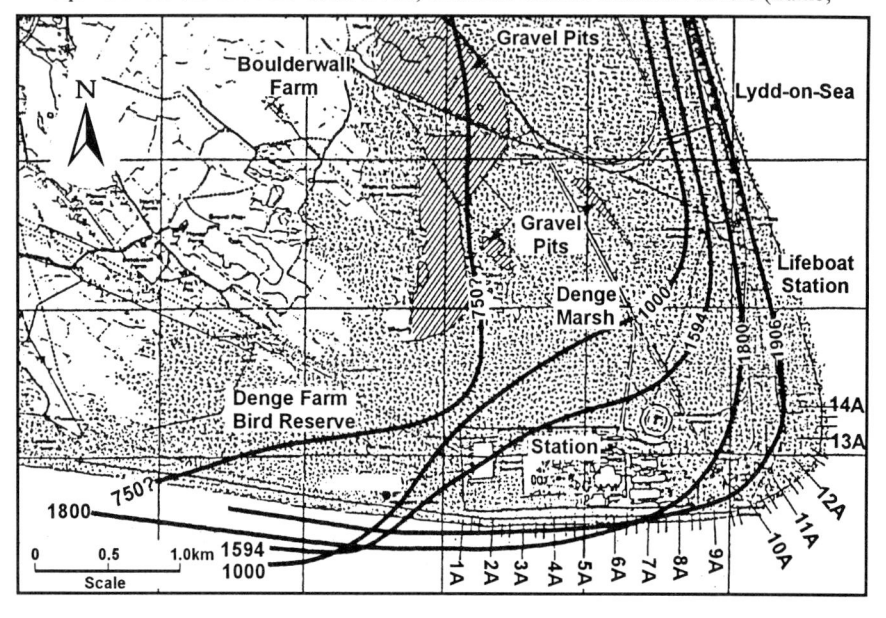

Figure 2 – Historical Coastline at Dungeness after Lewis (1932) and Lewis and Balchin (1940) and Annual Beach Survey Lines at the Station

1988). After 1300 years BP crest levels fall sharply, but then rise to their present day levels. On the accreting east coast, ridge levels reduce to landward.

There has also been considerable subsidence which, in the last 4000 years, is estimated to be between 1 and 2mm annually (Shennan, 1989; Long and Shennan, 1993; Long, 1994). The sea level rise is predicted to be about 0.4m by the year 2050 (Maddrell and Burren, 1990) giving a net relative rise of about 0.5m. The more recent positions of the coastline are shown on Figure 2, but the dating tends to be restricted to the last 1500 years, because of the lack of radiocarbon material. Lewis and Balchin's (1940) 750? shoreline was based on Ward's (1931) interpretation of two Saxon Charters. The 1600 shoreline is based on Pokers' 1617 map of Romney Marsh, while the 1800 shoreline is obtained from the 1794 Ordnance Survey. The 1990 BP coast is from recent radiocarbon dating (Maddrell et al., 1994), which suggests that Lewis' (1932) "C" line on Figure 1 represents the 1800 BP coast and the "B" line, the coast towards the end of the Neolithic.

Table 1 gives annual average rates of erosion and accretion for specified survey periods. The average long-term figures, weighted by the specified periods, show more accretion on the east coast than erosion on the south coast. This indicates that shingle was being supplied from the west, with no doubt some material moving through the area. The rate of erosion calculated for the period 1594 to 1816 will, however, reflect the accuracy's of the 1594 map. From 1816 to 1906 the erosion and accretion rates may be underestimated as not all the coast was covered. After 1906, there appears to have been more erosion than accretion and the volumes moved are decreasing as is the supply of shingle from the west, due to coastal works updrift and the development harbour mouth works in Rye. The ridge heights suggest that the most recent relative

Table 1
Coastal Changes

South Coast (erosion)		East Coast (Erosion)	
Dates	m^3/year	Dates	m^3/year
1594 – 1816	108,000	1816 – 1871	192,000
1816 – 1906	162,000	1871 – 1906	108,000
1906 – 1959	131,000	1906 – 1940	92,000
Long Term Average:	125,000	Long Term Average:	140,000

rise in sea level affected Dungeness from about the 15th century (Lewis and Balchin, 1940).

Hey (1967) found that the average rate of progradation of the beach face was about 5m annually which, with dips of 8 to 10°, gave an average rate of sedimentation of about 0.75m/year. From excavations for the Station he found that between 5 and 6 beds were laid down each year, which he attributed to periods of high spring tides, but not exceptional wave conditions. He also found the beach gradients were remarkably constant over 60 years, with erosion maintaining the gradient.

In the late 1950's the then Central Electric Generating Board (CEGB) chose the Ness as a nuclear power station site, positioning it next to deep (30m) fast flowing water. There are two stations, the oldest has two Magnox reactors, the other having two Gas Cooled reactors. Studies at that time by Sir William Halcrow and Partners (Halcrow, 1963) established that the rate of erosion of the southern shore was of the order of 1.1m annually, but could be up to 1.5m annually, with an annual rate of drift of about 125,000m^3. Various schemes for protecting the Station frontages from erosion, and thus flooding, were examined by Halcrow (1963) and one of recycling shingle from the accreting east face of the Ness was recommended. The proposed volume of shingle initially was 15,000m3 annually, rising to 20,000m3 with the construction of the second station.

Beach feeding has been continuous since 1965, concentrating mainly at the western, updrift, end of the site in the vicinity of Section 1A (see Figure 2). To the west the coast is an ecological and geomorphological Site of Special Scientific Interest (SSSI) and, while maintained at Jurys Gap some 8 km to the west, the area between has been left to erode. Thus, with this updrift erosion, the main recharge area became more and more out of balance, requiring increasing amounts of shingle to hold it. Recent studies recommended a policy of managed coastal retreat or set-back, which has been described in part by Maddrell et al. (1994) and Maddrell (1996).

1.2 Beach Feeding Strategy

1.2.1 General

Subsidence, sea level rise and the requirement to protect the Station from tsunami has meant an increased level of coast protection. The western corner of the site was inundated during construction (when the beach crest level was reduced) in 1960 and 1961 and locally in January and February 1983, February and November 1984, November 1989, January and February 1990 and August 1992. The need for flood protection was reflected in the recharge volumes which rose from 20,000m^3 originally to 70,000m^3 in 1992. The breastwork and two local groynes constructed as emergency protection in the early 1960's, are now buried.

The beach feeding scheme has operated for 32 years and Halcrow's annual reports eg Halcrow, 1970 and 1979, describe the work done during the previous season and recommend the quantities and positions for shingle feeding for the next. The volumes of beach feed material required are established by surveying specific sections of the beach during low spring tides at the end of the winter (see Figure 2). The recommended shingle quantities are then placed over the following winter i.e. from October to April.

The idea of recycling shingle for use as beach protection was unique at the time and has been the subject of regular review. The most recent review indicated that not only was this the most effective method of protection (costing less than the interest payment of a more formal and 'hard' methods of protection), but it had the lowest environmental impact.

Until 1972, the amount lost during the year was replaced at locations which had suffered the greatest erosion and was distributed in roughly equal proportions at a series of points. After 1972 a more efficient method of protecting the beach was adopted, placing most of the recharge updrift, allowing alongshore drift to distribute the shingle over the frontage. However, the erosive bay forming between the Station and Jurys Gap to the west, led to a gradual decrease in the shingle supply and the main point of beach feed between Sections 1A and 2A (see Figure 2), became more and more exposed. The beach feed increased to 70,000m^3, more than double the overall average annual volume during the entire project ie about 30,000m^3.

Initially, ground surveys were used to establish beach changes at the Station and borrow area, but in recent years it has been done by photogrammetry. The 1: 5000 photograph scale normally used allows individual beach levels to be determined to an accuracy of approximately $\pm$ 0.20 metres (assuming no errors in the control grid). There are 14 main survey sections each with three sub-sections (A, B and C; see Figure 2), 9 in front of the Station and 5 downdrift in the borrow area.

One of the major advantages of beach feeding has been its flexibility and the ability to respond quickly to potential problems and periods of intensive storms. Indeed, the records clearly demonstrate that the beach feeding has successfully stabilised the beach and the build up of the profiles has been sufficient to balance the increased losses in extreme years where the annual losses can be as much as 116,000m^3 (Maddrell, 1996). The increase in the total quantity of shingle in the beach profiles fronting the Station has also led to change in the "emergency" and "normal" profiles, conceived nearly 30 years ago. The original concept was that the "normal" profile should be the minimum cross-section of beach to be maintained, while the "emergency" profile represented a minimum profile which could lead to major overtopping and flooding, hence requiring immediate "emergency" works. Their profiles were altered in 1994 to account for the impact of a 1 in 10,000 year return period tsunami wave.

In addition to the regular surveys, fluorescent traces have been used (Russell, 1960). Overall the rates varied from 38,000m^3 to 192,000m^3 annually, averaging about 97,000m^3. In recent years the average annual rate at the station has increased to about 140,000m^3. More recently, "smart" pebbles have shown that transport follows the wave energy, although the larger pebbles move further, possibly because of their larger surface areas (Maddrell et al, 1994).

1.2.2 Past Investigations, Wind

In an effort to control costs and associated impacts of the beach feeding programme, Halcrow have for many years attempted to relate beach loss to local meteorological conditions (Maddrell, 1996). While this could never replace the surveys, it might allow their frequency to be reduced to once every two or three years. Initially the observed shingle losses were calculated with wind speeds exceeding 23m/s, ie. storm conditions. One of the difficulties in trying to establish any such correlation stemmed from the average annual gross loss representing a fairly thin layer over the surface of the total shingle volume contained within the sections. However, despite the drawbacks, the method of relating the number of hours the wind speed exceeds 23 m/s annually with the two year rolling mean of beach losses proved relatively successful.

However, since it was not entirely reliable, during the early 1980s this method of predicting beaches losses was abandoned.

1.2.3 Beach Loss Estimates

From the beginning it was appreciated that the incoming alongshore drift, which was about 120,000 m^3 per year, would slowly reduce. This was confirmed by later beach plan modelling, which showed the yearly gross shingle losses would increase on average by some 880 m^3/year. Analysis of the seven year rolling means of the actual losses in 1984 showed the annual rate of increase to be 540 m^3.

Since the start of beach feeding, the average gross rate of shingle loss up to 1992 was 29,500 m^3 annually. Annual gross loss from 1973 to 1978 was 24,000 m^3, rising to 25,000 m^3 from 1978 to 1983, 30,000 m^3 from 1983 to 1987 and to 42,000 m^3 from 1989 to 1992, i.e. the rate of loss was increasing rapidly in recent years.

1.2.4 Wave Energy and Beach Losses

Waves are both locally generated and Atlantic swell, with wind induced energy being some 75% and swell some 25%. The analysed data showed that, despite the wide scatter, there was clearly an increasing trend in alongshore transport due to increases in alongshore wave energy. Indeed, the difference between the predicted erosion rates using wave energy in an average year and the energy for more recent years, gives some 500 m^3/year increase in shingle loss, directly attributable to an overall increase in wave energy. The 380 m^3/year derived from subtracting this figure from the 880 m^3/year from the trend analysis of gross erosion may be attributed to the changes in alongshore drift due to coastal realignment updrift of the Station. (Maddrell, 1996).

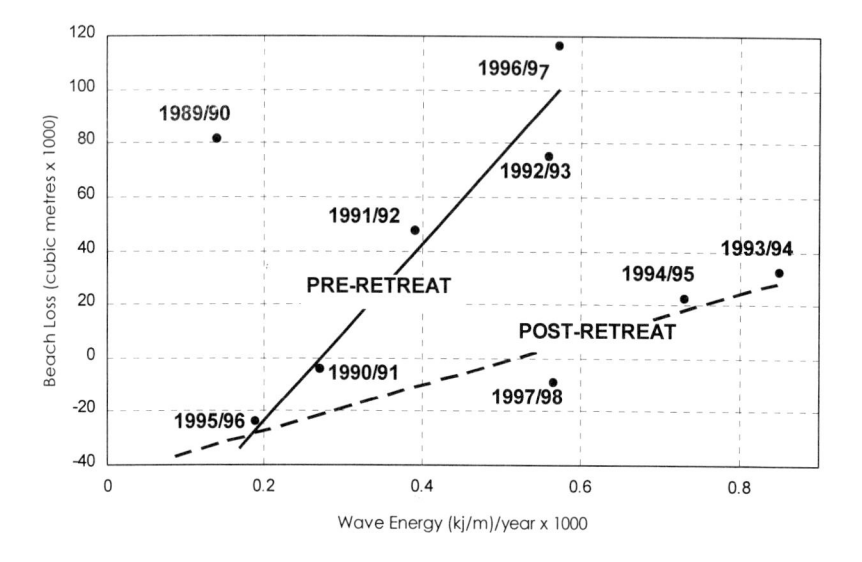

Figure 3 – Wave Energy and Beach Loss Quantities

While the trend lines show a clear correlation between increasing beach loss and wave energy, the comparison of individual years shows large deviations and therefore little predictive capability based on annual results. Figure 3 shows alongshore wave energy (wind waves and swell) plotted against gross beach loss for the winter beach feeding years from 1989 to 98 ie the combination of beach losses and beach feed. While there is a good fit for some years, the energy available for 1989 – 1990 was not, according to Hammond (1990), exceptional, yet the losses were high. This was mainly because the winter had four major storms, some with hurricane force winds, all of which coincided with spring tides. Indeed, Hey (1967) from his analysis of the excavated beach faces came to the conclusion that it is the coincidence of high water and waves and not wave energy alone, that controlled the beach face and ridges. It is apparent that, in general, a simple analysis of alongshore energy will tend to underpredict beach losses. Because of this and the occurrence of surges, the predicted and actual volumes can vary by as much as a factor of 2.

1.3 Optimisation of Beach Feeding

1992 saw, for the first time, a net loss of material from the licensed borrow area. Consequently, there was a review of the beach feeding strategy, which included a sediment transport model to predict drift rates and future changes in coastal morphology.

Three schemes were examined, namely:

- Continuing with the present policy;
- Supplying only the area near Section 4A, while allowing the western beach to evolve naturally.
- Constructing a long strong point at the eastern end to accumulate material and maintain the existing beach line

The model simulated fifty years for each of the schemes. Scheme 3 was discounted at an early stage because of the significant downdrift erosion and thus environmental damage associated with it. The estimated feed quantities for Schemes 1 and 2 are given in Table 2.

The present scheme of feeding near Section 4A (see Figure 2) shows considerable savings as:

- by moving the feed position downdrift, material stored within the bay between Section 4A and 1A and eventually to Jurys Gap is released,
- the beach length being protected was reduced,
- it released shingle to the east and
- the shortened length of the protected frontage has meant that only some 20,000m^3 is required annually.

The additional shingle taken in recent years has been used to raise and widen the coastal bund and to construct secondary bunds. The reason for increasing the dimensions of the main erodible western coastal bund is that where it has been overtopped updrift of the Station, if the bund is sufficiently wide, storm waves simply moves it on-mass inland, much in the same way as a sand dune moves. Indeed, Lewis (1932) notes that while waves erode the beach front, they also throw material onto the crest of the beach ridge. The evolution of similar protective shingle beach or ridge crests in Christchurch Bay has been described by Nicholls (1985). Powell (1989) describes the critical freeboard parameter for such beaches, which defines their stability.

1.4 Beach Feeding and its Alternatives

The economics of beach feeding at the Station were examined in detail in 1983 in connection with the studies for a proposed extension (Townend and Fleming, 1991) and the results can be seen in Table 2.

Table 2
Net Present Values of Coastal Protection Schemes at Dungeness, 1983 prices

Scheme		NPV, £x10⁶	
Life in years;		35	105
A:	Beach feeding	2.1	2.7
B:	Timber groynes and breastwork	4.8	5.7
C:	Mass concrete groynes and timber breastwork	5.7	6.5
D:	Armabrade steel piled groynes and timber breastwork	8.2	9.4
E:	Two strongpoints and initial feeding	4.3[a]	4.3[a]
F:	Revetment of armour units	15.3	15.3

[a] Plus cost of lee scour.

More recently the beach recharge was in addition compared with alternative forms of coastal protection, including alternatives such as pumping shingle (Bruun, 1990) from the borrow area rather than transporting it by lorry.

Even though the annual beach feed quantity in 1983 was only about 70% of that in 1992, the costs in Table 2 clearly demonstrate that, even with the strategy of maintaining recharge at Section 1, beach feeding was still the most economic form of coastal protection. The present policy of managed retreat or set-back significantly enhances the benefits.

1.5 Conclusions

- The decision, made in the 1960's to protect the Station by re-cycling the shingle was correct and farsighted.

- Beach feeding has been effective in preventing flooding at the Station.

- General increase in wave energy correlates well with increases in beach loss. Comparison of beach losses and wave energy for individual years do not.

- The coincidence of storms and high tidal levels have the greatest impact on beach losses.

- Ignoring the benefits of managed coastal retreat, beach feeding is still significantly less expensive than the other forms of coastal protection.

- The policy of managed retreat has lived up to its expectations, significantly reducing the quantities of recharge shingle required.

- Environmental impacts are confined mainly to the working areas.

2 APPLICATION ELSEWHERE

2.1 General

The low lying coastal land area, of England in particular, has increased significantly since Roman times and was in part encouraged by the fall in sea level during the mini ice age in the Middle ages. The fertile soils of the saltmarshes have been reclaimed almost continually since Roman times and while this meant a loss of saltmarshes, normally they continued to accrete seaward of the defences. Thus, while the coastal zone was squeezed, it gradually expanded again, and the Wash is a good example of this (see Figure 4 and Section 2.3).

While the old defences enclosed the newly reclaimed land, they were only relatively low earth embankments, being breached in storms and relieving pressure on adjacent defences. Today, however, the defences are higher and more sophisticated (essentially permanent reflective structures) and together with the rise in sea, they are putting pressure on the saltmarshes to seaward.

Despite the disastrous 1953 east coast flood, the Waverly Committee saw no reason to retreat from the then coastal defence lines, as the need for agricultural land after the Second World War was seen as paramount. However, now the need for such land is seen as less crucial (indeed productivity in some of these areas can only be sustained using chemicals). The cost of maintaining and repairing the defences is becoming difficult to justify against the economic value of the land and the true economic value of saltmarsh for grazing, breeding/nursery grounds for wildlife and fish is only now being fully appreciated. There is also increasing pressure from the insurance industry (Maddrell and Mounsey, 1996), who would like to see the limited resources available for coastal defences being directed at reducing risks to life and property in urban areas.

Breached defences (with full defences to landward) have the advantage that they provide the main defences with not only a wider foreshore area, but one which is vegetated, producing more drag and reducing wave energy. In addition, the remains of the old defences act as offshore breakwaters for many years. Indeed, studies on the Medway (French, 1999) have shown that even the rates of loss in the open marshes are greater over the same period of time as that for contemporary managed retreat sites. It is apparent that the breached sites have provided significant coastal protection to the hinterland for over 71 years, and have survived natural erosive processes more effectively than open marshes.

2.2 Blackwater Estuary
Two good examples of the changing attitudes to coastal defence can be seen on the Blackwater Estuary, at Tollesbury, which is a MAFF managed retreat experimental site and at Orplands, where the EA have breached as existing defence and allowed some 38 hectares to revert back to saltmarsh. The Orplands sea wall frontage covers 2km of coastline and was built in the C18 between 1740 and 1820, at the end of the mini ice age. The walls were small, hand dug, earth bunds protected by the saltmarsh infront and were cheap to build and cheap to maintain (Dixon and Weight, 1995).

The increase in tidal energy resulting from rising sea level has eroded the mudflats, foreshores and saltmarshes and Essex is losing 2% of its saltmarshes every year. Of the 50,000 hectares of marsh that Essex once had, only 4,500 hectares remain. However, some 40,000 have been enwalled and are now mainly agricultural land, with the remainder being eroded by natural forces.

The modern Orplands sea wall was more substantial, but had lost almost all of its protecting saltmarsh and its concrete blocks were being damaged by wave action. In

1994 it was shown that to reinstate the existing damage would cost £350k and to improve the wall over the next 20 years would cost an additional £250k to protect some 38 hectares of land. It would therefore have cost £16k per hectare to protect land, which had a market value of only £3.7k per hectare. Such a level of expenditure is uneconomic and is in breach of Treasury guidelines.

The EA had therefore no economic choice but to regenerate the old saltmarsh behind the existing Orplands sea wall and recreate a natural flood defence. The total scheme cost was about £80k, producing considerable savings and has been working successfully for over 4 years.

Subsequent monitoring of the Orplands site has shown that the recolonisation of the halophytic plants has been successful, with the vegetative cover increasing year by year in soils where the Redox potential is now the same as that for the external marshes. While there has been some minor erosion in the channels, there has been a tendency for accretion elsewhere. Fish and crustacea have returned, as well as worms and the site is regularly used by both breeding and wintering birds.

2.3 The Wash

The Wash is an important conservation area, supporting communities of wintering or migratory birds. The whole of the Wash has been designated as an SSSI, a Special Protection Area (SPA) and Ramsar site. In addition, the Wash is a candidate Marine Special Area of Conservation (cSAC). Recent studies have been carried out into the feasibility of managed retreat between Gibraltar Point and Hobhole Sluice, situated on the north-western shoreline of the Wash (see Figure 4).

As described in Section 2.1, large areas of the Wash have been reclaimed since Roman times. This has been due to the natural accretion and seaward advance of the saltmarshes, which have been mirrored by the reclamation (see Figure 4). However, not only are the present defences protecting the relatively recent reclamation areas, but much greater areas to landward over large parts of East Anglia. It is important therefore that their stability is maintained.

The proposed works would incorporate improvements to the inner banks to a 200 year standard, the construction of a new cross bank and the breaching of the outer bank, fronting Enclosures 7 and 8. The inner bank would be raised before the outer bank is breached, with material for the bank coming from various local sources.

A hydrodynamic study was undertaken to optimise the number and width of the breaches with the aim of restricting erosion both inside the site and to the exterior saltmarshes. Accretion also had to be limited as changes to the exterior SSSI/cSAC/SPA/Ramsar saltmarshes must be small in order not to damage their habitat. In addition, the drainage of the proposed new saltmarsh had to be developed.

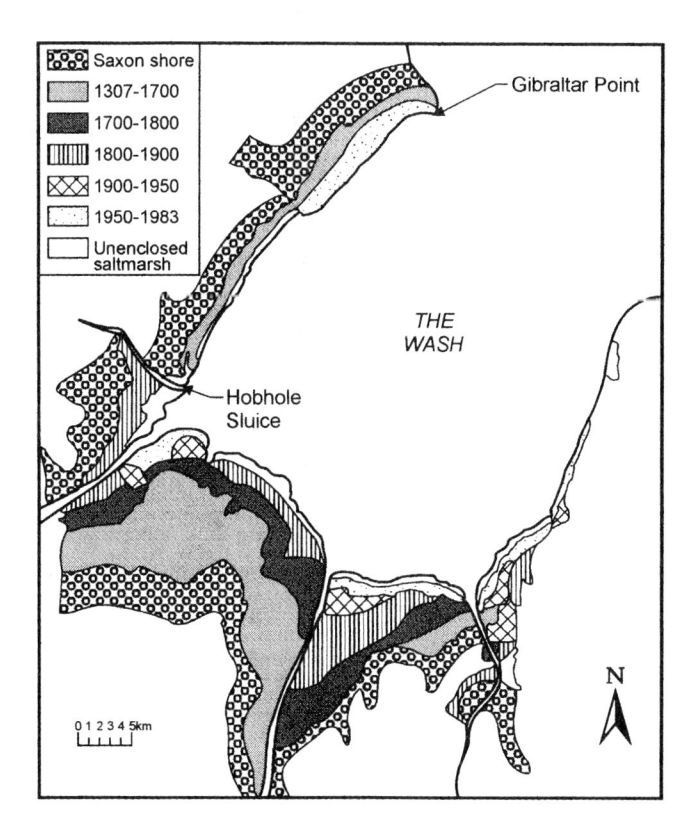

Figure 4 – Saltmarsh Reclamation in the Wash

Studies showed that increasing the number of breaches gave an improvement to the flow regime and reduced the potential accretion/erosion within the entrances and outer creek network. Increasing the width of the breach to 50m further improved the situation and wave modelling demonstrated that locally generated waves and waves from deep-water will not penetrate the managed realignment site under normal conditions. While accretion will occur, its rate will be gradual. The existing soke dykes will be maintained, with further secondary and tertiary channels developing as a result of natural processes as the land inside is in fact still the same level as or slightly higher than outside.

The site will lie between MHW and MHWS, with most being just above MHW and the plant community will include mid-lower to mid-upper marsh types. While the works may have environmental impacts, mitigation measures will be taken to reduce the potential impact of the works. These were reported on a separate Environmental Action Plan, which includes a monitoring programme.

The study recommended for the site:

- 3 breach points, each 50m wide, coincident with the existing exterior creek networks;
- re-establishment of the primary changes of the internal creek network;
- increasing the depth of the exterior creek network by 0.5m;
- maintaining existing soke drains;
- partially filling the existing open field ditches; and
- leave the remaining parts of the existing out bank.

The basic environment recommendations for construction were:

- no construction work 1 hour either side of a high tide;
- a walkover survey no more than one month before construction; and
- varying work areas if bank raising extended into the over-wintering period.

Due to its national and international nature conservation value, the recommended monitoring of certain environmental parameters included:

- a vegetation survey of the new bank and the inter-tidal area created by bank realignment assess colonisation and development of the saltmarsh;
- an invertebrate survey of the new inter-tidal area; and
- measurement of accretion and or erosion.

3   CONCLUSIONS

In the same way that, in the past, humans have been willing and able to exploit the changing coastal environment, now that those changes are working against us eg sea level rise, we should be prepared to abandon some coastal defences or, at the very least, critically review their value. The proposal is not one of wholesale abandonment, but of managed retreat.

The examples given in the paper show that managed retreat of:

- a shingle beach, provides an increased level of protection, at lower cost, to a perceived high risk site;

- healthy saltmarsh areas reduce the long term cost of flood defences;

- replacing farm land with new saltmarsh, reduces costs and protects the defences and the saltmarshes have a significant economic value in themselves.

REFERENCES

Bruun P., 1990. Bypassing Plants and Arrangement Prices on Transfers. In: Proceedings of the Skagen Symposium, Special Issue No. 9.

Greensmith, J. T. and Gutmanis, J. C., 1990. Aspects of the Late Holocene Depositional History of the Dungeness Area, Kent. Proc. Geol. Assoc., 101: 164 – 174.

Dixon A. M. and Weight R. S., 1995. Case Study at Orplands Sea Wall Blackwater Estuary, Essex, Proc. NRA Saltmarsh Conf.

Sir William Halcrow and Partners, 1963. Report on Coast Protection.

Sir William Halcrow and Partners, 1970. Beach Feeding Scheme, 5th Annual Report.

Sir William Halcrow and Partners, 1979. Beach Feeding Scheme, 14th Annual Report.

Hey, R. W., 1967. Sections in the Beach-Plain Deposits in Dungeness, Kent. Geol. Mag. Vol. 104, No 4: 361 – 374.

Hammond, J. M., 1990. The Strong Winds Experienced During the Late Winter of 1989/90 Over the UK: Historical perspectives. Meteorol. Mag., Vol. 119, No. 1419: 211 – 219.

Lamb, H. H, 1988. Weather, Climate and Human Affairs. Routledge, London and New York.

Lewis, W. V., 1932. The formation of Dungeness foreland. Geogr. J., 80: 309 – 324.

Lewis, W. V. and Balchin, W. G. V., 1940. Past sea-levels at Dungeness. Geogr. J., 96: 258 – 285.

Long, A. and Shennan, I., 1993. Holocene sea-level and crustal movements in southeast and northeast England, UK. Quat. Proc., 3: 15 – 19.

Long, A., 1994. Evolution of the south shore of Dungeness. Report for the Romney Marsh Research Trust.

Maddrell, R. J. and Burren K., 1990. Predicted Sea Level Changes and their Impact on Coastal Works Worldwide. In: 3rd Australian Port and Harbour Engineering Conference.

Maddrell, R. J., Osmond, W. O. and Bin, L., 1994. Review of some 30 years beach replenishment experience at Dungeness Nuclear Power Station, UK. Proc. 24th Int. Conf. Coastal Engineering, Vol. 3: 3548 – 3563.

Maddrell, R. J., 1996. Managed Coastal Retreat, Reducing Flood Risks and Protection Costs, Dungeness Nuclear Power Station, UK. Coastal Eng. 28: 1 – 15.

Nicholls, R. J., 1985. The Stability of the Shingle Beaches in the Eastern Half of Christchurch Bay, S. England. Ph.D. Thesis, University of Southampton.

Powell, K. A., 1989. The Dynamic Response of Shingle Beaches to Random Waves. Report CP19, H R Wallingford, UK.

Russell, R. C. H., 1960. The Use of Fluorescent Tracers for the Measurement of Littoral Drift. In: Proc. 7th Conf. Coastal Eng., The Hague.

Shennan, I., 1989. Holocene Crustal Movements and Sea-Level Changes in Great Britain. J. Quat. Sci., 4, 77 – 89.

Townend, I. H. and Fleming C., 1991. Beach Nourishment and Socio-Economic Aspects. Coastal Eng., 16(1): 115 – 127.

Ward, G., 1931. Saxon Lydd. Archaeol. Cantiana, 43: 39 – 47.

Pitfalls on the path to risk-based coastal management, and how to avoid them

JIM W. HALL
Lecturer, Department of Civil Engineering, University of Bristol, UK

SYNOPSIS
Risk-based approaches are becoming widely accepted by coastal managers. Now that the coastal management community in the UK is set on a risk-based course, it is timely to point out some potential pitfalls which risk methods may encounter if they not responsibly implemented. These pitfalls range from practical issues to do with the way risk methods are implemented and communicated, to deeply philosophical issues to do with the way models are used to make decisions in a complex socio-technical context. Provided it is recognised that risk methods provide only one perspective on uncertainty they can help to achieve more efficient coastal management decision-making.

1 RISK-BASED COASTAL MANAGEMENT
The aim of MAFF's flood and coastal defence policy[1] is:

To reduce risks to people and the developed and natural environment from flooding and coastal erosion by encouraging the provision of technically, environmentally and economically sound and sustainable defence measures.

Risk reduction therefore lies at the heart of all flood and coastal defence activities in the UK.

The concept of risk links the likelihood of an adverse event with the magnitude of the consequences of the event. Risk forms the basis of a rational way of making decisions under conditions of uncertainty. It is a concept that has attracted increasing attention in recent years from the point of view of project management, environmental impact and health and safety. Risk has also proved to be a useful concept in coastal management. Those involved with the coast are all too aware of the uncertainties in both the physical and organisational aspects of coastal management. Uncertainty is inherent in the random nature of hydraulic loads, in the complexity of structural, morphological and environmental responses, in the diversity of decision objectives and in the fallibility of the human system which implements flood and coastal defences. Some of these uncertainties can be effectively managed by employing risk-based methods.

A risk-based approach can be thought of as one which explicitly
- attempts to identify, in as complete a way as possible, yet to an appropriate level of detail, what may happen in future;
- assesses the possible consequences or impacts of future events;
- assesses, often in quantitative terms, the likelihood of different consequences or impacts;
- takes decisions on the basis of the assessment of likelihoods and consequences in order to manage the identified risks.

Coastal Management: Integrating science, engineering and management, Thomas Telford, London, 2000

It therefore links modelling with decision making, where the term 'modelling' is used in its most general sense.

2 DECISION MAKING

The aim of risk-based coastal management is to improve decision-making. It can do so by taking into account a wide range of possible outcomes and using the available information about what may happen in future to construct a probability distribution across those outcomes. The text book situation is illustrated in Table 1 where a set of decision options d_1, d_2,... d_l has been identified and analysed to establish how it is expected to perform in a set of future states of nature θ_1, θ_2,... θ_m which may materialise after the choice[2]. Depending on which state of nature in fact materialises, option d_i will yield one of m possible outcomes x_{i1}, x_{i2},... x_{im}. In a risk-based decision a model is used to generate the probabilities $p(\theta_1)$, $p(\theta_2)$,... $p(\theta_m)$ of occurrence of set of states of nature θ_1, θ_2,... θ_m. Suppose that $v(x_{ij})$ is the value associated with a given decision outcome x_{ij}. The decision-maker should chose the option that maximises $\sum_{j=1}^{m} v(x_{ij}) p(\theta_j)$, in other words the option which maximises expected value.

Table 1 General representation of choice problems

States of nature	θ_1	θ_2	...	θ_m
Probabilities	$p(\theta_1)$	$p(\theta_2)$		$p(\theta_m)$
d_1	x_{11}	x_{12}	...	x_{1m}
d_2	x_{21}	x_{22}	...	x_{2m}
Options .	.	.	...	.
.	.	.	...	.
.	.	.	...	.
d_l	x_{l1}	x_{l2}	...	x_{lm}

Given that
1. the states of nature θ_1, θ_2,... θ_m are an exhaustive set;
2. the model used to construct the probability distribution $p(\theta_1)$, $p(\theta_2)$,... $p(\theta_m)$ across the states of nature is a perfect one; and
3. the decision maker is neutral in their attitude to risk

then taking decisions that maximise expected value will in the long run be the best decision making strategy. The three provisos are strong ones and form the basis of some of the potential pitfalls of risk-based approaches that are discussed below.

3 PITFALLS AND HOW TO AVOID THEM
3.1 Uncertainty and probability

Risk provides a means of making decisions under conditions that are uncertain in the rather specific sense that there are a number of possible future outcomes but it is not known which outcome will materialise - it is only possible to assign a probability of occurrence to each of the possible outcomes. There are many situations in coastal engineering that are uncertain in this sense. For example it is not possible to predict the surge residual at a coastal site with certainty but, given tide data, it is possible to construct a probability distribution of surge water levels. However, there are other important aspects of uncertainty that are not as easy to capture in mathematical terms with probability theory. In many decision-making situations the available information is in the form of expert opinions, which are expressed in vague linguistic terms. There is therefore some imprecision of definition in the information available. This type of uncertainty is known as fuzziness. Information may also be conflicting or ambiguous. It will invariably be incomplete.

Some theorists argue that probability is a sufficiently expressive language to express all types of uncertainty[3]. Because it is by far the longest established and best-known mathematization of uncertainty, probability is the natural starting point for a rational theory of decision making. Under some circumstances decision makers seem to have little difficulty in expressing their uncertain knowledge and beliefs in terms of subjective probabilities, and these can provide useful evidence in a decision making situation (see section 3.8). But in situations characterised by vague, incomplete, ambiguous or conflicting information, the idealisations of probability theory will begin to look outlandish. Methods of handling uncertainty that are mathematical generalisations of probability theory are now quite well established and are beginning to be applied in coastal engineering[4]. When conducting a risk assessment a coastal manager should identify the types of uncertainty that are to be found in the problem at hand. Those aspects of the problem that can be thought of as being analogous to random processes can be expressed in probabilistic terms. Those aspects of uncertainty that are not amenable to probability theory should at least be identified and preferably be expressed in an appropriate syntax.

3.2 The meaning of risk

There is an email list server for risk analysts (riskanal@listserv.pnl.gov) where controversies over the meaning of terms like risk assessment, risk management, risk analysis and risk evaluation have smouldered for years and occasionally flared. The HSE has had difficulty in developing a common risk vocabulary and efforts by the ISO are controversial[5]. Introducing risk-based coastal management therefore faces the challenge that risk means very different things to different people, so it is important to state in what sense the term 'risk' is being used.

3.3 Risk perspectives

Amongst project managers there is inevitably a concern with risks *to the project*. The established ways of coping with these risks is to use techniques like risk registers to identify, prioritise and manage the risks. Meanwhile, clients, designers and contractors all have responsibility for risks *to the health and safety* of construction workers and members of the public. It is clear from MAFF's policy[1] (quoted above) that in coastal management there is another dimension to risk, which is the *risk to the developed and natural environment* posed by flooding and coastal erosion. Ultimately, the objectives of coastal project management, health and safety management and natural hazard management are convergent. Therefore, the aim should be for an integrated approach where decisions account for risks, objectives and constraints from all perspectives.

3.4 Weighing up values

The problem of conflicting values and objectives for coastal management is not necessarily a risk-based issue. It is fundamental to coastal management. However, problems of weighing up different values do come to the fore during risk-based decision making, based as it is on explicit statements of probability and consequences. For planning decisions (for example Shoreline Management Planning), and during early stages of strategy development and project appraisal it may not be possible or appropriate to express all of an option's attributes on a single (economic) scale. Under these circumstances so called multi-criteria or multi-attribute methods can be worthwhile. They can also be helpful in building consensus between project participants. If stakeholders in a coastal project hold fundamentally different values then this will become clear in a multi-criteria analysis.

Multi-criteria methods, as they are customarily applied in coastal management, do not take account of uncertainty in the scores that are applied to each option. This is potentially a serious shortcoming since it is at SMP stage or during pre-feasibility studies of project options that information relating to the future performance of options is at its most scarce and uncertainty is most acute. Some SMPs have proposed management options with little or no consideration of the uncertainty associated with the predictions of local and regional impacts. There are probabilistic ways of handling uncertainty in multi-attribute decision analysis[6]. However, a more straightforward approach is to place bounds on the values assigned to each of the options, rather than single values. If interval bounding methods are used it will not always be possible to identify a unique preferred option. However, at planning or pre-feasibility phase it will often be appropriate to recommend a small set of options for further study rather than specifying a unique preferred option.

Keeney and Raiffa[6] suggest that the set of criteria which are used in a multi-criteria choice should, as far as possible, be:
- *complete* so that it covers all the important aspects of the problem;
- *operational* so that is can be meaningfully be used in the analysis;
- *decomposable* so that aspects of the evaluation process can be simplified by breaking it down into parts;
- *non-redundant* so that double counting of impacts can be avoided;
- *minimal* so that the problem dimension is kept as small as possible.

Some attributes, for example sustainability, can be rather difficult to score, in which case it is advisable to decompose the attribute into measurable sub-attributes. However, when doing so it is important to avoid redundant sub-attributes.

3.5 Attitudes to risk

The approach to decision-making outlined in Section 2, which is based on maximising expected value, assumes that the decision maker is neutral in their attitude to risk. In other words they are neither risk averse nor risk prone. Various attitudes to risk can be included in theoretical approaches to decision making by using utility theory[2]. However, utility theory is not admissible to HM Treasury[7] because government investment decisions are generally taken from a risk neutral perspective. By adopting a risk neutral perspective the government aims to ensure that, over a large number of projects, the net benefit obtained for the tax revenue expended is maximised.

Nonetheless, decision-makers in flood and coastal defence may be particularly concerned about the possibility of extreme adverse consequences of their decisions. These extreme adverse consequences may amount to devastating flooding which results in loss of life and economic losses which are significant in national terms. In the face of these possible consequences decision-makers may, for legitimate reasons, display risk-averse attitudes. Rather than using utility theory, these situations can better be addressed by ensuring that all of the possible consequences, and the uncertainty associated with them, of extremely severe outcomes are highlighted. For example it may be appropriate to explicitly evaluate the risks of loss of life. If the risk of loss of life is found to be unacceptable according to criteria normally accepted elsewhere in government, it will be appropriate to constrain the decision options to those that satisfy the criteria for acceptable risk of loss of life.

3.6 Risk communication

A move to a risk-based approach to coastal management brings with it new challenges for communication between stakeholders in the coastal system. It is clear from the analysis conducted in the UK[4] that engineers have difficulty in communicating even quite basic probabilistic concepts of flood risk to the populations at risk, at a time of growing public concern yet widespread confusion about risk issues in general. Reliability calculations are seldom transparent to stakeholders who have an interest in the outcome of those calculations. The danger of adopting authoritarian technocratic approaches that are not transparent is that the co-operation from politicians and the public, which is necessary to implement flood defence works, will not be forthcoming[8]. Therefore, if coastal engineers are not to undermine their own best endeavours on society's behalf for safety, efficiency and sustainability, they need to recognise the social construction of risk and develop appropriate mechanisms for communicating risk issues.

3.7 Models of variable dependability

To make a risk-based decision requires some probabilistic model of the future performance of decision options. The model will consist of some prediction of what may happen in future (the states of nature $\theta_1, \theta_2, \ldots \theta_m$) (see Table 1) and information about the relative likelihoods of future scenarios, expressed as a probability distribution $(p(\theta_1), p(\theta_2), \ldots p(\theta_m))$. The models employed may be the subjective 'mental models' we all carry with us, or the more 'objective' models available through shared theory. However, a model is by definition not reality. It is an abstraction, which is only partial and is incomplete. Incompleteness can be due to a number of different causes[9]:

- That which cannot be foreseen - this includes the existence of phenomena which were previously unknown.
- That which is foreseeable but is
  (a) ignored in error - these effects should be taken into account but are missed by mistake;
  (b) ignored by choice - these effects or this data is explicitly considered to be unimportant; or
  (c) ignored because of lack of resource - these effects are too complex to model at this time or the data required is too costly to collect.

The results of a risk analysis will reflect the incompleteness of the models used.

A risk analysis will usually involve assembling information from several sources and using a wide variety of models. Analysis of the project appraisal and design of a managed set back project on the East Coast of the UK identified 117 different data collection and analysis processes[10]. Naturally some of these processes will be more dependable than others. Yet model results expressed purely in terms of a probability distribution provide no information about the dependability of the models used. Without some understanding of the dependability of input distributions and response functions, the data generated by probabilistic methods can be difficult to interpret and in the wrong hands quite misleading. It is therefore essential to express the dependability of the models that have been employed and reflect that dependability in subsequent decision making. Probabilistic statements should be supported by a logical audit trail of through the evidence upon which they are based.

3.8 Dependable use of expert judgement

Whilst in a risk assessment one may hope to use probabilities derived from quantitative or statistical analysis there are situations in flood and coastal engineering where this is not possible, in which case recourse will have to be made to expert judgement. An element of expert judgement is inevitable in any analysis. Even a seemingly quantitative statistical analysis is based on judgements of what data to admit in the data set used for the analysis and of the applicability to the data set to the problem at hand. Nonetheless, there is an ample body of literature to demonstrate that expert judgements, and specifically subjective judgements of probabilities, can be prone to bias[11]. Some good practice guidelines for minimising bias in expert judgements of probabilities are outlined as follows.

3.8.1 Precisely define the events to which the expert is being asked to attach probabilities.
Before eliciting a probability, the analyst should ensure that the expert is attaching an accurate meaning to the event for which a probability is being elicited. For example, is the expert considering the probability of one flood defence breach *only* in the next n years, or one *or more* breaches? The distinction between the cumulative probabilities over a number of years and annual probabilities should be made clear.

3.8.2 Structure the problem logically with the help of event trees.
In complex situations it will be very difficult for experts to handle the multiplicity of factors which determine the probability of system failure. It is often preferable therefore to decompose the problem using event trees and ask experts for judgements of scenarios that make up the event tree.

3.8.3 Check the expert testimony for inconsistencies.
By logically structuring the elicitation of subjective probabilities and obtaining several judgements relating to different aspects of a problem it is possible to check for inconsistencies in the expert testimony and, where they do exist, work with the expert to develop a more coherent set of probabilities.

3.8.4 Make use or peer review.
Expert judgements should be subjected to critical review by the expert's peer group. Approaches to conducting peer review vary from informal discussions to more formal mechanisms for eliciting collective judgements such as the Delphi method[12].

3.8.5 Use any available quantitative data to inform the expert judgement.
In many studies, even if a full probabilistic analysis is not possible without recourse to subjective probabilities, it is often the case that there is some statistical data, which could be incorporated in the analysis. For example it is usually possible to obtain reasonably dependable statistics for the occurrence of extreme storms which are associated with initiating defence failure. This statistical information can be used at one node in an event tree and the expert can then be asked for the conditional probability of failure *given that* a specified storm or number of storms has occurred.

3.8.6 Document all of the evidence upon which the expert judgement was based.
Expert judgements of probabilities are based on evidence which will range from the tacit knowledge of the expert to specific analysis, data and historic evidence relating to site in question. The expert judgement should as far as possible be made transparent by documenting the sources of evidence and the process by which expert judgements of probability have been obtained.

3.8.7 Use expert judgements of bounds on probabilities or 'most likely', 'best' and 'worst' estimates rather than point estimates.

Experts in flood and coastal engineering will willingly acknowledge the uncertainty that is inherent in probabilistic predictions. The expert's uncertainty should be expressed in the probabilistic predictions in terms of a probability distribution (often based on three estimates: 'most likely', 'best' and 'worst'). Alternatively the bounds on the estimate may be treated separately in the analysis to provided an indication of the sensitivity of the final decision to the expert's uncertainty.

3.9 Basic health warnings of probability and statistics
Some simple checks will help to avoid elementary blunders in probabilistic analysis.

3.9.1 Violating basic probability theory
1. The probabilities of an exhaustive set of events should some to unity.
2. The conditional probabilities of an exhaustive set of conditional events should sum to the marginal, or in other words, if $B_1...B_n$ is an exhaustive set of events:

$$P(A) = \sum_{i=1}^{n} P(A | B_i).$$

If either of these rules is violated then a mistake has been made somewhere in the analysis.

3.9.2 Probabilities and frequencies
An analysis of risk requires a *probability* of a clearly defined event. Extreme value statistics may also be expressed in terms of frequency or return period. Whilst for extremely rare events annual probability and annual frequency converge they are fundamentally different quantities and it is bad practice to use them interchangeably.

3.9.3 Dependency
In complex systems there will often be strong dependencies between different elements or processes. This is true of physical phenomena, for example wave heights and water levels at the coast, which often display some correlation. It is also true of variables such as project cost where there may be strong dependency between the construction cost of different project elements. Numerous elements of flood warning and emergency response systems can be influenced by common organisational issues, so it is unrealistic to assume that the failure rates of different elements of the system are independent. The presence of dependencies between different system variables will have a major influence on risk calculations.

3.10 Limits to predictability
It seems that there may be limits to the predictability of some coastal processes in the long-term and on a large spatial scale, when results obtained by integrating small-scale physical processes in short time steps become questionable. De Vriend[13] argues that long term trends in coastal evolution are "a weak residual of a very 'noisy' signal of short term variability". The presence of dynamic scale interactions, through which small-scale phenomena can lead to large-scale effects, leads de Vriend to conclude that the predictability of coastal behaviour is restricted.

Even when models of coastal change do provide dependable results there can be practical limits on their use for probabilistic purposes. Models may be too slow for practical numerical convolution of probability distributions. Doubtless computer power will make it more practical to use models in a probabilistic mode. However, it seems likely that the most complex hydrodynamic and sediment transport models will always be close to the limit of available computing power. The probabilistic modeller will have to make do with models that are an order less complex if they are to be able to numerically integrate probability distributions.

3.11 Soft engineering

The established methods of quantitative risk analysis in engineering originated in the process and nuclear industries. These are safety-critical systems that can be thought of as comprising of a large number of discrete components connected together in some complex but clearly identifiable way. Statistical failure rates of the components are well documented. Early work on reliability of coastal structures endeavoured to transfer techniques developed in these industries, notably fault tree analysis, to coastal engineering. Event tree and cause-consequence analysis has proved to be more applicable, but still only for fairly uniform defence types. It is not surprising that the most comprehensive applications have been for analysis of Dutch dikes.

Soft coastal defence systems present much more of a challenge. Failure mechanisms interact and are characterised by progressive cycles of decay rather than by discrete failure events, so a fault tree approach is less applicable. Indeed failure itself is rather difficult to define and certainly does not conform well to the precise failure threshold adopted in reliability theory. Behaviour is often strongly three-dimensional, so analysis that involves analysing several cross-sections, which are assumed to be joined in series or parallel, is hardly applicable. Temporal change is also a major feature of soft engineering systems.

3.12 Robustness and resilience as well as efficiency

An optimum solution as identified by a risk-based approach may be vulnerable to unforeseen events that were not included in the risk analysis. Much of the value of a resilient coastline[14] is its capacity to cope with the unforeseen and unpredictable, in particular climate change. Collingridge[15] suggests that unforeseen outcomes in socio-technical systems can be controlled by:
- monitoring,
- reducing the cost of failure by managing vulnerability and exposure,
- reducing the corrective response time,
- reducing the cost of remedy,
- keeping one's options open by adopting flexible solutions and enhancing variety.

The intention of these strategies is to improve the robustness of a plan or design in the face of events that are not predictable by the models available to the coastal manager at the moment of decision making. These benefits cannot be fully evaluated in probabilistic terms. It may be necessary to weigh the apparent loss of expected value involved in adopting a robust decision, against the apparent loss of flexibility in adopting the 'optimal' decision. The balance will tend to favour robustness in conditions of high uncertainty.

3.13 Observational methods

Many of the shortcomings of risk-based decision-making originate from a reliance on predictive models. It is possible to reduce over-reliance on predictive models by adopting observational methods. The observational method was introduced in geotechnical engineering by Ralf Peck[16] where, in common with coastal engineering, the conditions actually encountered on a project often depart significantly from those predicted by site investigation and modelling. Peck suggested that rather then designing for the worst case it can be more economical to design for the most likely condition and then put monitoring in place and be ready to change the design as the actual conditions on site materialise.

Observational methods already form the basis of soft engineering techniques like beach nourishment. Renourishment operations are designed in response to feedback from monitoring. The need for predictive models is not eliminated by observational methods. A model is still required in order to make decisions, but by actively monitoring and responding to monitoring information it is possible to cope when the system starts to depart from the model predictions. Feedback from monitoring can also be used to improve the predictive models, so they become 'learning' models in a dynamic system.

The observational method fits in well with the general view of coastal management as being a process of controlling a dynamic system. The coastal manager is continuously receiving monitoring signals, and responding to those signals in order to achieve some target response. It is the close linkage between physical systems and coastal management systems that makes soft coastal engineering a resilient and potentially sustainable strategy. Yet that close linkage also limits the extent to which the performance of the system can be predicted, so makes it particularly challenging from a risk assessment point of view.

4 CONCLUSIONS

Risk-based approaches provide a formal way of taking account of some of the uncertainties that confront the coastal manager. They enable a wide range of future conditions, from the every-day to the extreme, to be accounted for in a decision. Whilst in the past uncertainty was managed in an intuitive or implicit way, for example by employing factors of safety, risk methods help to externalise uncertainty, making decisions more transparent and potentially more efficient. However, risk-based coastal management is no panacea. In industries where risk assessment has been in use for much longer than in coastal engineering, the pitfalls of risk-based methods are becoming increasingly recognised[17]. Even in the idealised domain of decision theory, risk-based decision making will only pay off in the long run, and is no safeguard against things going wrong.

Many of the problems associated with risk methods are generic issues to do with the use of models. Models, by definition, are imperfect representations of reality, and some models are much more dependable than others. Probabilistic methods and risk assessment can be so elegant and engaging that the naïve decision-maker is beguiled into what Herbert Simon referred to as bounded rationality[18], whereby the existence of significant phenomena which are not represented in the available models are neglected altogether.

Foremost amongst the problematic issues which do not fit in comfortably in a risk analysis are organisational, political and values issues. Engineering failures are as much a sociological phenomenon as a technological one and flooding disasters are no exception[19]. Risk methods may help to externalise the values issues at the root of the conflicts that can embroil flood and coast defence policy and projects, but they do not provide a mechanism for resolving them. To do so requires consensus-building processes.

Risk methods provide but one perspective on a complex phenomenon. Their naïve adoption will result in over-reliance on fallible predictive models and the misguided assumption that every type of uncertainty can be modelled in probabilistic terms. However, risk methods can help to improve decision-making if their pitfalls are recognised and they are used as but one tool in the management of the uncertainty.

5 ACKNOWLEDGEMENTS

The research upon which this paper is based was part-funded by HR Wallingford and the Environment Agency. Valuable comments on a draft of this paper were received from Ian Meadowcroft in the Environment Agency's National Centre for Risk Analysis and Options Appraisal. Aspects of the paper were inspired by the author's involvement as one of the co-authors of MAFF's forthcoming *Flood and Coastal Defence Project Appraisal Guidance: Approaches to Risk*, but the views expressed are the author's own and no endorsement by the MAFF is implied.

6 REFERENCES

[1] MAFF Strategy for Flood and Coastal Defence in England and Wales. *MAFF publication PB1471*. 1993.
[2] LINDLEY, D.V. *Making Decision*. London: Wiley, 1971.
[3] LINDLEY, D.V. Scoring rules and the inevitability of probability. *International Statistical Review*, Vol.50 (1982) pp.1-26
[4] HALL, J.W., DAVIS, J.P. and BLOCKLEY, D.I. Uncertainty analysis of coastal projects, in *Proc. of the 26th Int. Conf. on Coastal Eng.*, Copenhagen, June 22-26, 1998.
[5] ISO/TMB Risk management terminology – Guidelines for use in standards. *ISO/TMB Second Working Draft on Risk Management*. April 1999.
[6] KEENEY, R. L. and RAIFFA, H. *Decisions With Multiple Objectives: Preferences and Value Tradeoffs*. New York: Wiley, 1976.
[7] HM TREASURY *Appraisal and evaluation in central government: Treasury guidance*. London: The Stationery Office, 1997.
[8] BECK, U. *Risk Society: Towards a New Modernity*. London: Sage, 1992.
[9] DAVIS, J.P. and BLOCKLEY, D.I., 1996. On modelling uncertainty, in *Proc. of the 2nd Int. Conf. on Hydroinformatics*, Zurich, Sept., 9-13, 1996.
[10] HALL, J.W. *Uncertainty Management for Coastal Defence Systems*. PhD thesis, University of Bristol, 1999.
[11] KAHNEMAN, D., SLOVIC P. and TVERSKY A. *Judgement Under Uncertainty, Heuristics and Biases*. Cambridge: Cambridge University Press, 1982.
[12] ROBERDS, W.J. Methods for developing defensible subjective probability assignments. *Transportation Research Record*, 1288 (1990) pp.183-190.
[13] DE VRIEND, H.J. Mathematical modelling and large-scale coastal behaviour. *J. Hydraulic Research*, Vol. 29, No. 6 (1991) pp.727-753.
[14] NICHOLLS, R.J. and BRANSON, J. Coastal resilience and planning for an uncertain future: an introduction. *Geographical Journal*, Vol.164, No.3 (1998) pp.255-258.
[15] COLLINGRIDGE, D. *The Social Control of Technology*. Milton Keynes: The Open University Press, 1980.
[16] PECK, R.B. Advantages and limitations of the observational method in applied geomechanics. *Geotechnique*, Vol.19, No.2 (1969) pp.171-187.
[17] VAN BREUGEL, K. Beyond the risk concept, in *Proc. 7th Int. Conf. on Structural Safety and Reliability*, Kyoto, 24-28 Nov. 1997.
[18] SIMON, H.A. *The Shape of Automation for Men and Management*. New York: Harper and Row, 1965.
[19] BYE, P and HORNER, M. Easter 1998 Floods - Final assessment by the independent review team. *Report to the Board of the Environment Agency*. 30 Sept. 1998.

Managing Risk and Uncertainty in the Design and Construction of the Minehead Sea Defences

WEST, M.S; CAPORILLI, M & SEDGWICK, P
Mouchel Consulting Ltd, West Byfleet, UK

INTRODUCTION
The situation at Minehead
Minehead is situated on the Somerset coast and is exposed to both a large tidal range and an exposed wave climate, which can include both the residue of Atlantic swell waves and locally generated waves over a fetch of many kilometres. The town itself was a busy fishing and trading port in the 17th century. The harbour and beaches of the area are now used by tourists who provide a significant source of local income. As the town developed, a series of sea defences were built to varying designs and standards of service. The frontage concerned is heterogeneous, containing shingle, cobbles and sand in front of a mixture of vertical masonry walls, stepped revetment and a wave return wall. Many of the seawalls constructed along Quay Street and the Esplanade at the beginning of the century were showing signs of fatigue. At the eastern end of the frontage, the shingle ridge was inadequate as a defence. Prior to the present scheme, flooding of the town occurred frequently, with significant recent events in January 1996, February 1990 and December 1981. The area protected by the defences is low-lying with the land to the south of the holiday centre reclaimed from marshland. Recent development has placed commercial property including light industrial; retail and leisure facilities in the area most at risk from flooding.

A Brief History of the Scheme
Severe flooding during a storm in December 1981 prompted the Wessex Water Authority to identify and review the options for improving the standard of defence at Minehead during the first half of 1982. Mouchel was first appointed to appraise the options for reducing the risk from flooding due to overtopping of the seawall in May 1992. Under several appointments from that date, the scheme has been gradually developed as indicated in the programme shown in figure 1. The final scheme as approved for grant aid by the Ministry of Agriculture Fisheries and Food is shown in figure 2. The programme shows the conceptual design phase leading to the production of the Engineer's Report, required to secure funding, as well as the public exhibition and subsequent construction periods. It is anticipated that completion of the final element, the nourishment contract, will be achieved during the summer of 1999. This should enable this year's holidaymakers to enjoy the amenity beach without the interference of construction traffic and for the residents and businesses in Minehead to be protected from flooding prior to the autumn 1999 storms.

The Aims of this Paper
This paper has been written as a case study, which, by virtue of its timing and certain characteristics of the frontage, illustrates well the application of both traditional and modern approaches to risk management in design and construction. Whilst the two have been separated for convenience of reporting, the involvement of all interested parties over a long

200 COASTAL MANAGEMENT

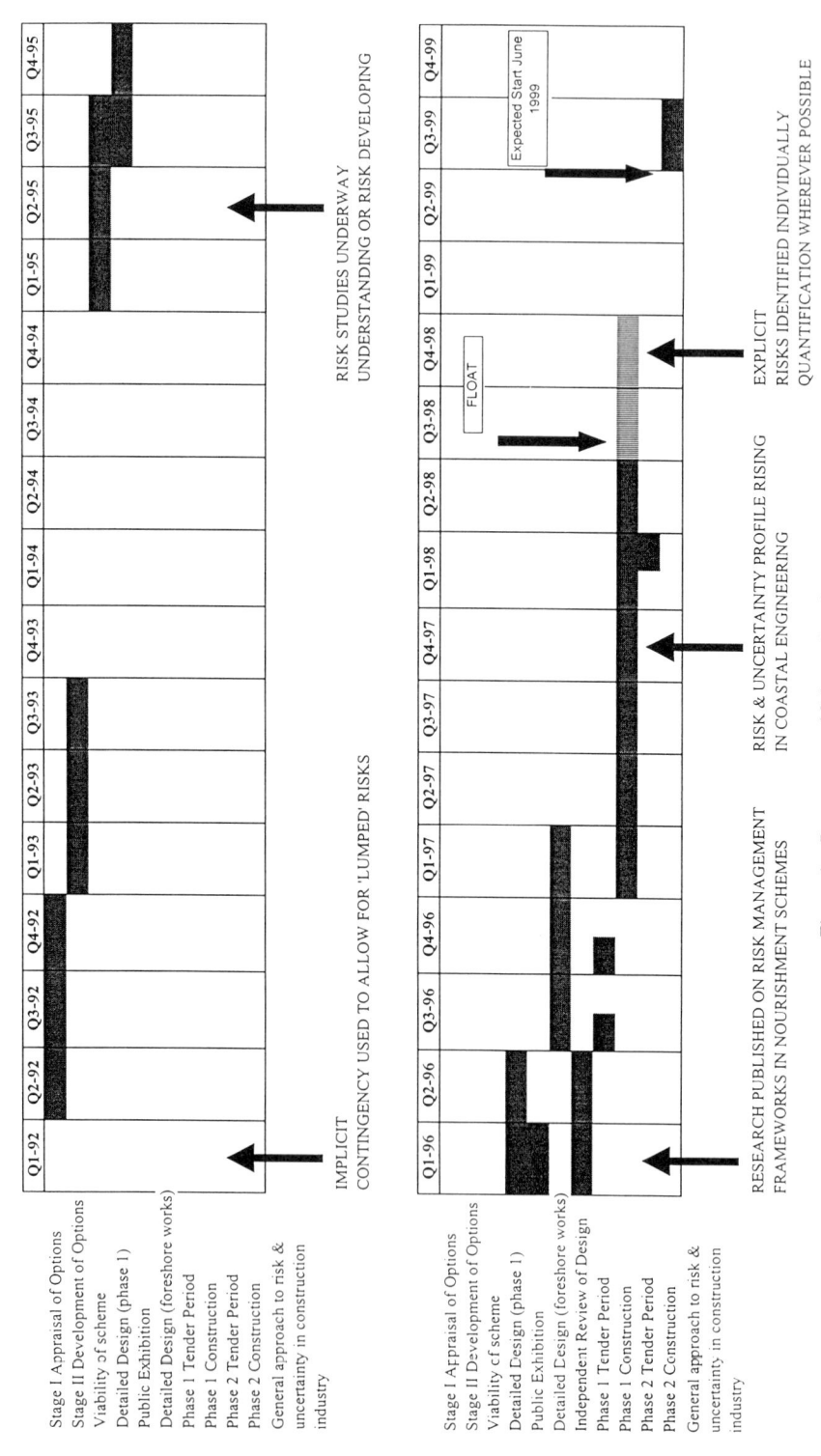

Figure 1 - Programme of Scheme Development

PAPER 20 : WEST, CAPORILLI AND SEDGWICK 201

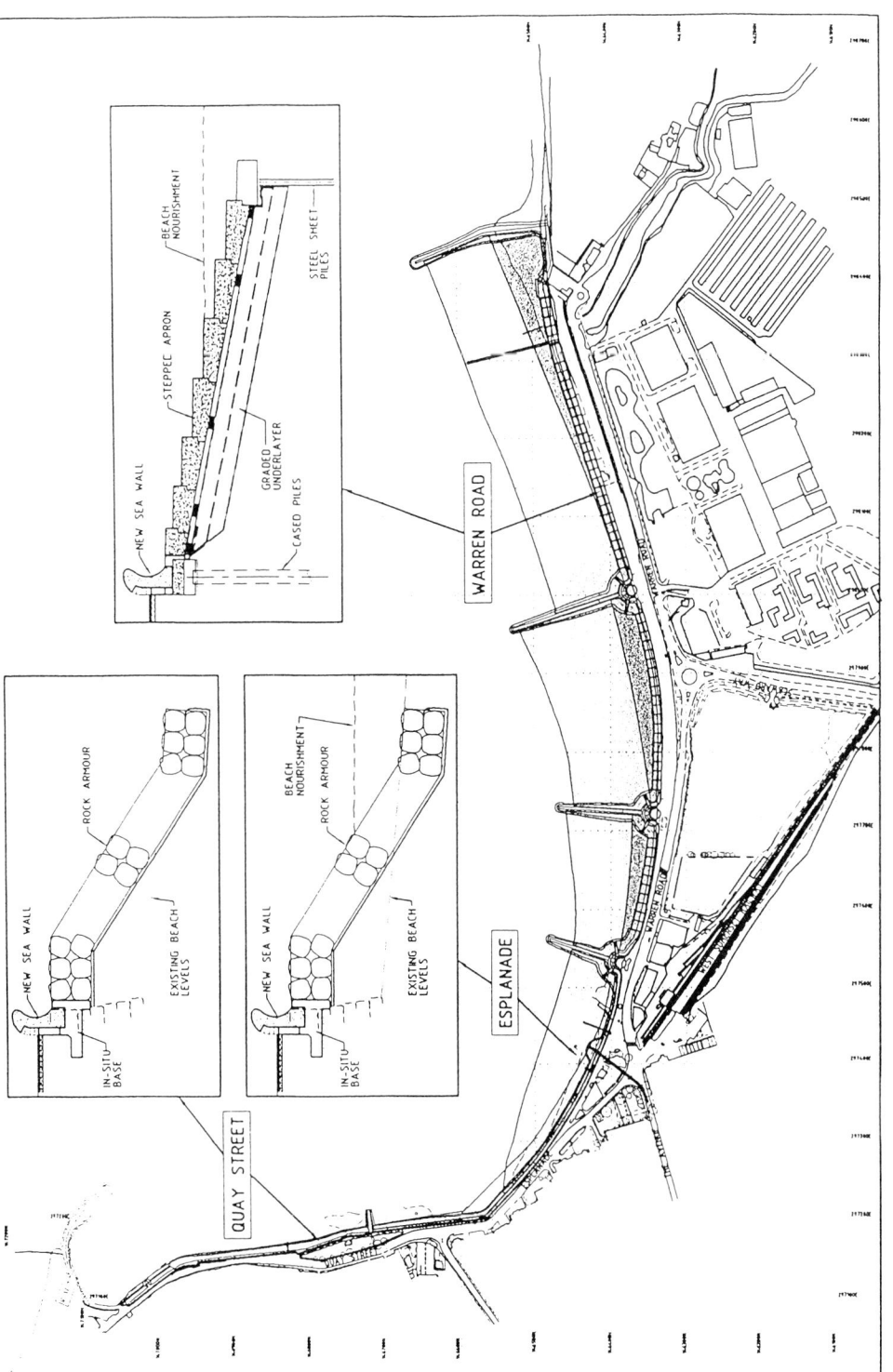

FIGURE 2 - GENERAL ARRANGEMENT OF APPROVED SCHEME

period of time has ensured that construction issues were considered during the design process with the objective of managing the uncertainties within a range of complex constraints including financial, planning and the inevitable environmental constraints imposed by the wave and water level climate.

The case study has briefly reviewed the modern approach to risk management with reference to the Minehead scheme. Then a number of risk and uncertainty issues have been illustrated by descriptions of parts of the design process and the evolution of the procurement strategy. An emphasis has been placed on those issues that illustrate messages of wider relevance wherever possible.

RISK & UNCERTAINTY
Definitions
Risk is both defined and measured as the probability of a specific unwanted event and its unwanted, consequential loss. The quantitative impact of risk is the product of the probability of the unwanted event occurring and the consequential loss incurred. Uncertainty underpins much of the need for risk assessment and the management of risk in both design and construction. There is uncertainty associated with predicting the occurrence of many natural hazards (eg. design wave height) as well as economic variables (eg. material prices). 'In all of these, there is no way to predict the magnitude of the hazard in the future' (Simm & Cruickshank, 1998).

The Approach to Risk & Uncertainty in Design & Construction
The evolution of robust techniques for dealing with risk and uncertainty in coastal engineering and their adoption in very recent years has been a key issue at Minehead. In the early stage of the scheme's development, risk management was applied instinctively. Specific risks such as that posed by the wave for which the scheme was designed (the 'design wave') were quantified however many risks remained implicit and were only managed by 'judgement informed by experience' (Simm & Cruickshank, 1998). As the scheme has been designed and is now under construction, so the approach to risk management has evolved. Many of the implicit, previously unquantified risks, are now recognised explicitly on the understanding that describing them formally should make them easier to manage and will certainly make it harder for them to be completely overlooked.

A Holistic View of the Risks & Uncertainties at Minehead
In accordance with the methodology proposed by Floyd (1996), a rapid assessment of the uncertainty in the beach nourishment element of the scheme at Minehead was made. The results are shown in table 1.

	Attribute	Score	Weight	Weighted score	Comment
1.1	Beach conditions	4	1.5	6	Complex bay shape Various grain sizes exist
1.2	Survey work	1	1	1	Reliable surveys
2	Wind conditions (offshore)	2	3	6	
3	Wave conditions (nearshore)	3	1	3	No wave data from nearshore
4.1	Raw fill	3	1	3	

4.2	Processing	1	0.5	0.5	
5.1	Model used	3	4	12	Validated packages used Not calibrated for Minehead
5.2	Storm occurrence	3	3	9	
5.3	Storm design	3	1	3	
6.1	Grain sizes	3	2.5	7.5	
6.2	Beach profile	2	1.5	3	
			Total	**54**	

Table1. Assessment of Nourishment Scheme Uncertainty

This assessment is based on a framework for uncertainty quantification and the scores obtained should not be considered precise or beyond challenge. The overview does however illustrate particular features of the Minehead scheme and highlight those areas where uncertainty remains.

According to the indicative scale of uncertainty proposed by Floyd (1996), Minehead is characteristic of a scheme with 'considerable uncertainty involved and the predictions of performance could be out by ±50% or more'. This finding is consistent with the judgement made by those engineers involved in the design of the beach retaining structures and of the beach nourishment. For consulting engineers, such a reinforcement of the opinions of individuals may have significant consequences both for the way the client views the design (and the designers) as well as in the commercial risk inferred for the designers (i.e. the professional indemnity liability).

The features of the beach at Minehead, which led to the largest uncertainties, are the complex bay shape, which has a radius of curvature of approximately 1.5-2km, and the variety of beach sediments which occur close to the nourishment area (from fine mud to large cobbles). During the design process the lack of nearshore wave data was a problem and, although numerical models were used to supplement the poor available data, these could not be calibrated using measured data. It was not possible to calibrate either the wave transformation or the sediment transport model. As a result, a key conclusion from the design of the nourishment was that beach performance monitoring post construction is vital to quantify the actual rate of sand loss from the groyne bays. This should enable the maintenance programme to be optimised so that losses are minimised. The measured (unavoidable) losses can than be used as the basis for planning a periodic renourishment strategy.

RISK IN DESIGN
Virtually all of the decisions made during the design process were either motivated by a desire to manage risk or had implications for the total risk and uncertainty associated with scheme performance or construction. In the section that follows, some of the risk issues that arose during the design process are described.

Meteorological and Tidal Conditions
Vital to the accurate calculation of future performance, as well as key to determining bad weather risks during construction and the risk of structural failure of the various elements of the scheme, was the quantification of the wave and water level climate at the site. A review of available data indicated that the site was not well covered by existing databases. No long-term wave statistics were identified from the area and it was necessary to develop them using

wave hindcasting based on wind speed and directions from the Bristol Channel. Long-term water levels were obtained from Avonmouth and adjusted to the site. An extreme value analysis was conducted on both sets of data to obtain independent statistics for water levels and waves. A joint probability analysis produced a range of water levels with an associated wave height for each of a number of return periods.

To provide the wave and water levels required for design at the location of the structures, the offshore wave climate was transformed to points within Minehead Bay using a wave ray model. This included wave refraction and shoaling. The near shore wave statistics produced were adjusted to allow for the wave height limiting affect of breaking. Unfortunately no wave data existed from within Minehead Bay so it was not possible to rigorously calibrate or verify the near shore wave data used in the design. Such a fundamental uncertainty in the design conditions resulted in a 'conservative' approach being taken when calculating forces either on the seawall structures or when looking at the potential mobility of nourished material. This conservatism was most clearly visible in the rounding-up of wave heights and the rejection of potential nourishment sources that were not demonstrably coarser than the native beach sand.

Calibration of Overtopping Performance
Before the future overtopping performance of the scheme could be predicted, a model was required which could accurately reproduce the behaviour of the existing seawall. No quantitative data were available concerning the existing overtopping situation and the only source of information was anecdotal. A physical model was constructed to establish benchmark performance data with which designed improvements could be compared. The expense of physical modelling was such that it was not possible to simulate every scheme element under every combination of water level and wave condition. Careful selection of model test parameters was necessary to ensure that the limited budget was applied to maximise the reduction in uncertainty achieved by the modelling exercise. Following this rationale, only the most exposed length of the frontage (Warren Road) was simulated in the physical model and empirical techniques used to predict overtopping along Quay Street and The Esplanade.

The physical model was constructed in a flume that allowed the processes of shoaling and breaking to be included but was unable to simulate refraction. Since the latter causes wave attenuation, the absence of refraction will have resulted in slightly higher waves in the model than in the prototype situation. This is an example of the conservatism referred to above.

Single events with return periods of 1:1, 1:10 and 1:100 years were selected and tested in the physical model. The overtopping discharge measured in the model was used to calibrate an analytical model of overtopping. The latter was then applied to each of the water level and wave condition combinations to identify the worst case for overtopping. The worst case was not that with the highest water level and it depended on the configuration of the seawall under test. Figure 3 shows how the worst case for the existing seawall was experienced with a water level of 6.70mODN and Hs of 3.00m for a 1 in 100 year return period event, whereas for the same return period and the proposed scheme, it occurred with a still water level of 6.75mODN and Hs of 2.78m.

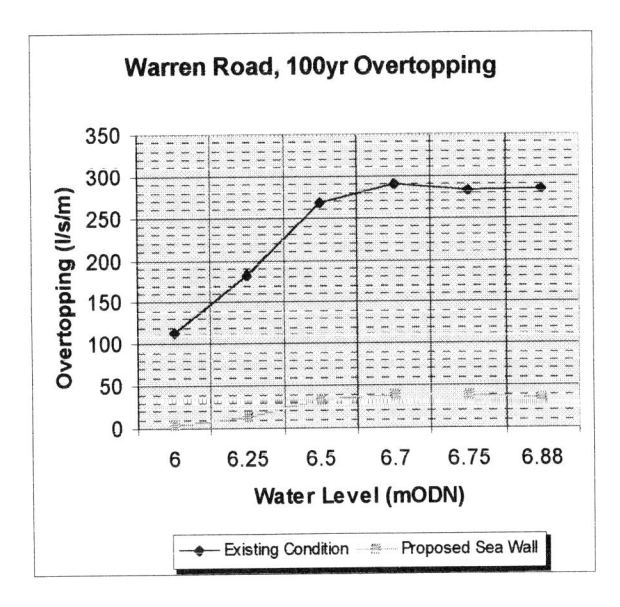

Figure 3 – Variation in Overtopping Discharge

Whilst the lack of a reliably calibrated model of overtopping in the existing situation has resulted in some uncertainty in the prediction of performance, the combined use of physical and numerical techniques and the emphasis on comparing relative improvements in overtopping. The methodology has reduced the uncertainty by as much as possible given the financial constraints which are always present and which, in the design process, inevitably mean than judgement and experience are required to interpret the results from modelling and other design calculations.

RISK IN PROCUREMENT
Constraints with an Impact on Risk Management

The approach taken to managing risk in the procurement of the scheme was determined by constraints imposed by stakeholders outside the immediate project team of Client-Consultant-Contractor. Budgetary constraints imposed on the Environment Agency Flood Defence Committee required that the Contractor be paid over four fiscal years. Environmental and planning constraints sought to ensure that the bulk materials, especially rock, would not be transported to the site by road.

The performance of the existing defences was inadequate and there was a real risk of failure either by excessive overtopping of the seawall or via a breach of the shingle ridge at the eastern end of the frontage. Pressure to act quickly resulted in a decision to procure the hard defence elements of the scheme whilst the detailed design of the beach nourishment and beach management structures was still underway.

Initial Procurement Strategy

The best way to accommodate the constraints was to split the construction into two phases. Phase 1 would involve a general civil engineering contractor who would demolish the existing seawall, build the replacement and add the stepped apron. Most of the works involved concrete construction and the design was produced with pre-cast unit construction in-mind. This was intended to reduce risk by increasing the consistency of the mix and encouraging investment in high quality formwork for the casting of repeat sections. Pre-cast units would also reduce the risk of excessive or unpredictable downtime due to adverse tidal and storm surge conditions. Only a small quantity of rock, essential at the location of the Access/Viewing Points (AVPs) was included in phase 1. It was planned that a subsequent phase 2 would include the importing of the bulk of the rock for the revetment and groynes.

It was envisaged that the rock would need to be sourced from outside the region and would therefore require delivery to site by sea. This work could then be combined with the delivery of the beach nourishment, also by sea. Planning constraints required the contractor to avoid the tourist season and also restricted working hours to minimise noise disturbance to residents. To minimise the risk of flooding through a section of the seawall which was under construction, the contractor was only permitted to work on individual 25metre sections of wall with a maximum of 50m of seawall unprotected at any time. It was expected that these constraints would limit the rate of progress and that payments could be made in accordance with the budgetary constraints.

Tender Evaluation
On examination of the tenders for phase 1 it became apparent that the cost of importing a small volume of rock was disproportionately high and that investment in a railhead at Minehead would allow materials to be imported by land but without an adverse increase in vehicle traffic. At this time the design of the rock elements of the scheme had been completed. This had shown that local rock from the Mendips was acceptable and that it could be imported using the new railhead.

Revised Strategy
A re-examination of the risk of the project indicated that including all of the rock work in phase 1 would reduce the risk of non-performance (i.e. overtopping). Delivery of rock to the site by rail would avoid the risk of adverse weather at sea interrupting supply and would also contribute more to the local community since a railhead would be established at the railway station. The phase 2 works would comprise solely the provision and placement of the beach nourishment. These works would be let such that the partially completed scheme was only exposed to one winter storm period.

The major obstacle to adopting the alternative strategy was the financial constraint of spreading the capital investment over a four-year period. The solution was a financing arrangement enabling the works to proceed without any artificial restriction on the rate of spend but requiring the contractor to fund the working capital which the client would repay, with interest, once funds were available. This solution enabled significant capital cost savings of over £700,000 to be made.

The financing approach was novel in the area of sea defences at the time and required special approval from both the Environment Agency and Ministry of Agriculture, Fisheries and Food. A strong business case was prepared which clearly described the costs and benefits of achieving early completion and which demonstrated the risk and uncertainty involved in both procurement strategies.

Re-tender
The two lowest tenderers were invited to re-tender for the works with an alternative that provided for the rock revetment and groynes to be included in this phase of construction. The tender documents included a financing clause, which allowed the contractor to set out the interest terms for any sums due, but not paid through the normal certification. The new tenders confirmed the benefits of the revised approach. Following evaluation and negotiation, the phase 1 works were awarded to Tarmac Civil Engineering (TCL) in a contract with a duration of 104 weeks.

Construction of Phase 1
The design of the phase 1 works, which made maximum use of pre-cast concrete and other repetitive elements (rock armouring, steel sheet piling), the contractors 'conveyer belt' approach to construction, and the spirit of co-operation which existed on-site, enabled the phase 1 works to be completed some six months ahead of programme. This reduced the Health & Safety risk to the general public as the beach was free from construction equipment during the summer of 1998. It also reduced the risk of claims for compensation by those businesses in Minehead which are dependent on tourists for their livelihood.

During the autumn of 1997, when it became apparent that the phase 1 works were likely to be completed early, it was decided that the phase 2 works should tendered sooner than planned so that they could be implemented as soon as the phase 1 works were complete.

Phase 2 Tender
Once again the financial constraint imposed by the Environment Agency's capital programme was a constraint. It was hoped that advantage could be taken of the early completion of phase 1 and that the phase 2 works could proceed in the autumn of 1998 however The date when funds for the phase 2 works would be made available was uncertain.

The strategy adopted to manage this risk involved writing the tender documents for phase 2 in such a way that the flexibility to undertake the works during any of three financial years (98/99, 99/00, 00/01) was provided.

Following receipt of tenders, a business case was made for the early implementation of the phase 2 works, setting out the advantages and benefits to be gained from early completion of the scheme as a whole. These advantages were set against the financial impact resulting from the risks associated with delayed completion. A robust improvement in the benefit/cost ratio was demonstrated. Beach response to storm waves during the winter of 1998/99 gave rise to with localised lowering of the beach adjacent to the new defences. This highlighted the risks of damage to the partially completed scheme and of the likely requirement to procure higher volumes of nourishment material during phase 2 to replace any losses as a result of erosion in the interim period.

Before the award of the contract a formal risk workshop under the direction of a facilitator was arranged with participants from other Environment Agency offices and other consultants in addition to the project manager, Mouchel project team and contractor. This process enabled remaining risks to be identified and classified with appropriate mitigation or contingency actions identified. The resulting risk register has been reviewed regularly since its preparation in autumn 1998.
At the time of writing, award of the phase 2 works has been made with completion in the summer of 1999.

CONCLUSIONS

The handling of risk in the design of the scheme have been shown to depend crucially on the availability of information to quantify the wave and water level climates at the site. A methodology involving both numerical and physical modelling was employed to reduce the risks but some uncertainty could not be avoided. A key recommendation of the design report was that the performance of the nourishment be monitored following placement so that a refined strategy for managing the beach could be prepared as the level of uncertainty was reduced.

Risk in the procurement of the scheme has benefited from a flexible approach and the adoption of novel financing methods, specifically partnerships in which contractors bring some financial resources to the project.

The Minehead scheme was designed and is being constructed during a period of rapid change in the way risk is managed in coastal engineering especially. The challenge faced by project team members has been to include the latest thinking whilst retaining continuity and momentum in the programme. There is clear evidence of both Environment Agency's and the Treasury's willingness to adopt new ideas where they can be shown to be of benefit.

Some of the language and techniques now being applied to manage risk have a lot in common with those used in the Private Finance Initiative. It is therefore likely that language such as 'Best Value' and 'Public Private Partnerships' will also be adopted in coastal engineering in the months and years to come.

REFERENCES

Floyd, P.J (1996) 'Assessing the Risk of Beach Recharge Schemes', National Rivers Authority, R&D Note 508.

Simm, J & Cruickshank, I (eds)(1998) 'Construction Risk in Coastal Engineering', THOMAS TELFORD.

ACKNOWLEDGEMENTS

The authors wish to express their appreciation of the efforts made by the Environment Agency's project team, led by John Taberham, throughout the design and construction of the scheme and for permission to publish this paper.

Does coastal management require a European Directive? The advantages and disadvantages of a non-statutory approach.

S. JEWELL, H. ROBERTS & R. McINNES
Isle of Wight Centre for the Coastal Environment, Isle of Wight Council, Newport, UK.

1. INTRODUCTION

Coastal zone management has been on both the national and European political agendas for at least 25 years (Huggett, 1996) as a result of their increasing use by many different industries and sectors of society, often with competing interests. In addition, an estimated 200 million (30%) Europeans (out of a total of 680 million) live within 50 km of coastal waters.

The ultimate aim of coastal management is to ensure sustainable use. In an area where there are so many organisations with responsibilities it was widely recognised early in this decade that the sectoral approach to managing the coast that had been adopted to date would not fulfil this aim. A multi-disciplinary approach was needed to stimulate co-operation and co-ordination of activities in order to bring about consensus-building in managing competing activities. This requirement was recognised by both national government and the European Commission (EC). In the UK, a voluntary process was advocated and reinforced in The Department of Environment's publication Coastal Zone Management - *Towards Best Practice* which stated *"The way forward must be integrated management to achieve common goals delivered locally through **voluntary partnerships**."* (DoE, 1996).

The EC also recognised the need for integrated management and initiated a Demonstration Programme on Integrated Coastal Zone Management (ICZM) in 1996 involving 35 projects located throughout the member states of the European Union (EU). One of the original objectives of this programme was to inform the EC as to the action it should take at a European level with regard to ICZM. The Commission originally thought that this might result in the production of a Directive on coastal management; as it receives results from the thematic experts[1] and initial feedback from the projects, it is now erring towards a more flexible, non-statutory approach encouraging local, voluntary initiatives.

Over the past two years the Isle of Wight Centre for the Coastal Environment, within the Isle of Wight Council, has been undertaking one of these projects[2]. As part of the project the team have examined the advantages and disadvantages of such non-statutory approaches through its 'audit' of coastal zone management along the central south coast of England.

---

[1] Six thematic experts were commissioned to draw together findings from the 35 projects. The reports can be found on http://europa.eu.int/comm/dg11/iczm/home.htm.
[2] funded under the auspices of the EU-LIFE (L'Instrument Financière de l'Environnement) Environment and TERRA programmes.

Coastal Management: Integrating science, engineering and management, Thomas Telford, London, 2000

After a brief overview of the key developments of ICZM in both the UK and at European level, this paper uses local examples to demonstrate the successes of voluntary initiatives within the study area, the benefits this brings to integrated and collaborative arrangements and the resulting improved understanding between organisations in the coastal zone.

However, the paper also outlines the disadvantages of this approach when working within a national and European Union policy vacuum. This paper discusses these aspects to determine if there is a middle road, combining the stability of the statutory planning process with the flexibility of the voluntary, non-statutory, bottom-up approach. It highlights the pivotal role that local authorities have come to play in ICZM in the UK and examines what contribution the EU could play in moving towards sustainable use of the UK's, and indeed the EU's, coasts.

2. BRIEF HISTORY OF ICZM AT EUROPEAN UNION AND UK LEVEL

2.1 Influence of the European Union

The Council of Europe began promoting integrated coastal planning to help ensure wildlife conservation in coastal areas in the early 1970s. Since that time there have 00been repeated calls for action both at the EU and member state level. Key initiatives are listed in Table 1. An early milestone was the European Coastal Charter produced by the Conference on Peripheral Maritime Regions (CPMR) in 1981. Debate continued into the next decade and in 1991 at the European Union for Coastal Conservation (EUCC) Congress called for a European strategy for integrated planning and management of all European coastal zones. At the beginning of this decade there was also a call for a Directive on ICZM, but this was quashed due to a lack of agreement between Directorates General (DGs) within the EC.

By 1995, in response to increasing pressures on the coastal zone, the Commission adopted a communication on ICZM which reviewed the main features and the state of the EU coastal zones. The Council of Environment Ministers adopted a resolution calling for the production of a European Community strategy for ICZM, however, progress towards the production of this strategy has been slow. To speed up the process and progress coastal management a Demonstration Programme was launched in 1996[3], with a view to showing the benefits of improved information and co-ordinated mechanisms for the implementation of sustainable development and identifying the need for further action at EU and other levels.

Demonstration projects have taken a variety of approaches and these experiences are being reported to the Commission. Using this information and that of the thematic experts, in May 1999 the Commission launched two consultation documents (European Commission 1999a and 1999b). These documents mention the importance of bottom-up voluntary initiatives as one of a number of instruments for delivering ICZM, but stress that "coastal zone management is **not** effective if it is not supported by all levels of administration..." (p.13, European Commission). Conferences are now being held with key organisations in each member state to obtain feedback on how the Commission should proceed. It is likely that a statement about the way forward will be released at the end of 1999.

---

[3] Reflecting recognition for the need for closer co-operation and co-ordination of actions the Demonstration Programme is a joint action between three DGs (DG XI – Environment; DGXIV – Fisheries; and DG XVI – Regional Policy and Cohesion) under the financial instruments of LIFE and TERRA.

Table 1: Key European Milestones on ICZM

Year	Milestone
1973	Council of Europe Resolution (73) 29 on the protection of coastal areas urged Governments to compile inventories of coastal resources, and to promote integrated coastal planning and ensure wildlife conservation in coastal areas.
1975 & 78	European Commission studies on development, conservation and integrated management of coastal areas in the EC.
1981	European Coastal Charter adopted by Conference of Peripheral Maritime Regions of the EC.
1982	Resolution by European Parliament supporting the Coastal Charter.
1985	European coast a major subject of the 4th Council of Europe Ministerial meeting.
1986	EC Communication to Council of Ministers on integrated planning of coastal areas (COM/86/571) – concluded regions had not fully applied the Charter.
1991	Communication from European Workshops on CZM requesting EC to take urgent action to promote integrated CZM.
1991	EUCC Conference at Scheveningen. EC requested to "prepare a Community strategy for integrated CZM which will provide a framework for conservation and sustainable use."
1992	Council of the European Communities Resolution (92/C 59/01) invited the Commission to prepare a Community strategy on this matter and incorporate the initiative into the Fifth EAP.
1992	Fifth EAP included support for a EU strategy on CZM and noted that a high priority should be given to the environmental needs of coastal zones through inter alia better co-ordination between relevant EC policies and policies at the EC, national and regional levels. Also stated that the framework for integrated management plans, at appropriate levels should be prepared by 1998.
1993	European Environment Bureau submit a memorandum on CZM to the EC to develop a legal instrument to support, guide and encourage progress on CZM.
1993	EUCC conference in Marathon recommends that the EC prepare a CZM Directive.
1993	IUCN/CNPPA conference drafts Action Plan for Protected Area in Europe including support for an international framework for CZM and a European Community Directive on the coast.
1993	First draft of a Handbook on economic development and environmental protection in coastal areas, prepared as part of the EC ENVIREG programme.
1994	Council of Ministers meeting to discuss CZM strategy.
1995	Commission Communication on ICZM with objective of testing co-operation models for ICZM, to provide technical results to foster dialogue between EU institutions and coastal stakeholders.
1996	Demonstration Programme, involving three DGs (XI,XIV & XVI) funding 35 projects and six thematic expert, launched to identify appropriate measures to remedy the situation of deteriorating coastal zones of the EU.
1998	Coastal zones incorporated into European Spatial Development Perspective as an Environmentally Sensitive Area with specific conference in Gothenburg.
1999	Consultation on findings of Demonstration Programme through hosting of conference to launch discussion document in each participating member state and wide dissemination of document. Consultation period May – September.
1999	Final document with recommendation of how to take forward the European ICZM strategy due December 1999.

2.2 National Influence

In the absence of clear direction from the EU, most member states have adopted non-statutory approaches to the ICZM process; the UK is no different in this regard. The Demonstration Programme has provided resources to take forward local, voluntary initiatives in a way that otherwise would not have been possible.

In terms of the national agenda, at the beginning of this decade there was a strong push for the UK Government to take some action on coastal zone management. In 1992 the House of Commons Select Committee Inquiry on Coast Protection and Planning recommended **a voluntary, top-down, bottom-up approach** to delivering ICZM and envisaged a **"cascade" of plans and strategies** operating at different levels. However, in its response the

Government was not in favour of producing national or regional coastal plans, but felt that "...*local authorities and other agencies will be expected to work together on a **voluntary** basis, informed by an understanding of national and regional issues provided by **national policy statements** and by [other national and regional] bodies...*"(DoE 1993).

The DoE have produced a number of documents in relation to coastal planning policy (listed in Table 2 below). 'Coastal Zone Management - Towards Best Practice' (DoE, 1996) provides key points for good practice based on examples of local voluntary, non-statutory initiatives underway at that time. In this document the Government remained focused on supporting 'the sectoral approach to regulation on the coast whilst encouraging a multi-agency, *voluntary* approach to coastal zone management' (p12 DoE, 1996). However, with the exception of Planning Policy Guidance 20: Coastal Planning (DoE, 1992), which deals primarily with the statutory planning process, the national policy statements mentioned in the Government's response to the Select Committee report have not been forthcoming for more holistic management.

A number of voluntary initiatives have since been started resulting in a range of management plans and strategies. The Isle of Wight's project Demonstration Project, has examined such mechanisms along the central south coast of England, looking at both the statutory and non-statutory tools. The following sections explain the role of the voluntary approach in delivering ICZM and highlights the successes and problems encountered.

Table 2: Key milestones in the national guidance of ICZM

Year	Milestone
1992	House of Commons Select Committee Inquiry on Coast Protection and Planning.
1992	Production of 'Policy Planning Guidance 20: Coastal Planning'
1993	Publication of 'Managing the Coast: A review of Coastal Management Plans in England and Wales and the powers supporting them'
1993	Publication of 'Coastal planning and Management: A review'
1995	Publication of 'Policy Guidelines for the Coast'
1996	Publication of 'Coastal Zone Management Towards Best Practice'
1998	House of Commons Select Committee Inquiry on Coastal Defence.

3. EXAMPLES OF VOLUNTARY APPROACHES WITHIN THE STUDY AREA – SUCCESSES AND PROBLEMS

3.1 Development of Plans and Strategies

Statutory plans are produced by the local planning authorities. In England, they are a product of the legal process set out in the Town and Country Planning Act 1990, and advice set out in Planning Policy Guidance Note 12 "Development Plans and Regional Planning Guidance". Their scope is restricted by legislation to land use policy or proposals only, so they cannot fulfil an all embracing function. However although not devoted solely to the coastal zone they deal with many of the development issues found in this area. The process requires full participation and consultation with the public and aims to provide policies and proposals that cover a ten to fifteen year life-span. Their main drawback are that they only have jurisdiction to the mean low water mark and they are not flexible enough to reflect the dynamic nature of the coastal zone.

A number of holistic, non-statutory management plans have been developed to complement the statutory planning system. They vary in their geographic scope from regional level such as the 'Strategic Guidance for the Solent' (Solent Forum, 1997), through to plans for designated sites. These non-statutory management plans have increasingly opened up important opportunities for co-ordinating the range of bodies and interest groups involved in the coast, and have provided a mechanism for resolving conflicts. They are generally voluntary documents which are designed to inform, clarify and expand upon issues which are recognised in the coastal zone but which are not adequately addressed through the existing statutory plans.

Initiating the debate
The first dedicated coastal zone management plans within the study area, such as the 'Hampshire Strategy for the Coast' (Hampshire County Council, 1991) and the South Wight Borough Council's 'Management Strategy for the Coastal Zone' (McInnes, 1994), tended to be a statement of policies and views from one organisation, usually the local authority. They showed the basis on which management practice could be developed, rather than actually developing policies for the coast in the more participatory fashion that is now accepted as good practice. However, these early non-statutory plans were felt to be worthwhile because they showed a clear statement of intent by the local authorities and looked at issues across the land-sea interface and planning-management divisions. In many ways these strategies were seen as invaluable in initiating the debate on coastal issues in the area, especially as at that time such issues were only just coming on the national agenda. The growing number of voluntary plans over the past decade has been recognised by most parties as a positive move towards improving the management of the coast. But the shear number of them can cause confusion as to their actual value and use in delivering ICZM.

Informing the statutory process
As mentioned above non-statutory plans also assist in developing consensus through participation in issues that do not receive adequate attention through the statutory planning process. An example of this is when Hampshire's Strategy for the Coast tackled the issue of provision of new moorings for yachts in the county. Through a process of consultation involving all the relevant stakeholders, a consensus was achieved that any new provision of moorings should be restricted to built up areas and should not affect town or seascapes. This view was incorporated in Hampshire County Council's Structure Plan four years later with very little opposition. It is highly unlikely that this acceptance would have been achieved without the process that arose from the voluntary coastal strategy.

Success is not universal
The success of this example is not universal. Due to the lack of clear policy advice at national level, there is still no structured national framework within which non-statutory plans can sit. Consequently, there is no clear guidance on the way in which these plans should be implemented. The Solent, with its multitude of issues and users swiftly moved on from the Hampshire Strategy to embrace a more holistic approach involving the wide range of users which meet as the Solent Forum. 'The Strategic Guidance for the Solent' (Solent Forum, 1997),developed after many rounds of consultation was one of the first plans to be produced that moves beyond addressing issues that are primarily the concern of local authorities and takes a more integrated approach to managing the coast.

As a voluntary document there is more flexibility to incorporate a wider range of policies. Opportunities exist to use the policies within the Strategic Guidance to inform statutory development plans on issues that lie outside the jurisdiction of the planning system, or would not receive the attention required to develop a consensus through participation of all key stakeholders. This work has relied on the commitment and perseverance of one or two individuals who had the ability to see the wider picture. But they were, and still are operating in a national policy vacuum. It is only good-will that allows these non-statutory policies to be taken into account in the statutory process. As a result this approach is not uniform along the coast.

For example, in neighbouring West Sussex, laudable work has been undertaken, including the publication of a Coastal Strategy by the County Council (W. Sussex County Council, 1994) and a number of coastal zone management plans by the district authorities (Arun District Council, 1993, Chichester District Council, 1995). In the absence of national policy advice, political will and support has been lacking and the process has stagnated. The move has not yet been made to the more holistic approach that takes in the ideals and aspirations of all stakeholders that has been seen to be so beneficial in the Solent.

3.2 The Establishment of Coastal Groups and Fora

It is widely recognised that a key facet of ICZM is effective communication and co-operation. In common with the rest of the UK, significant progress towards this goal has been achieved in the study area through the establishment of a number of groups and fora. As with plans, these can be either sectoral or holistic and vary in their geographical scope. For example, the coastal group SCOPAC (Standing Conference on Problems Associated with the Coastline) was set up in 1986 to improve co-operation between local authority coastal engineers and related groups along a 400 km stretch of the south coast of England. SCOPAC primarily deals with issues relating to coastal defence and has contributed significantly to the more integrated approach to shoreline management seen today. Membership of this group is voluntary and regular attendance by all local authorities and county councils within the area, as well as MAFF, Environment Agency and English Nature is a testament to its success.

Building on the achievements of SCOPAC and recognition nationally that a more holistic approach to managing the coast was necessary, The Solent Forum was set up in 1992 involving over 70 members and representing over 50 organisations. Membership ranges from conservation organisations such as the Royal Society for the Protection of Birds to large, private commercial enterprises such as Associated British Ports; all statutory authorities are also represented. The Solent Forum is hosted by Hampshire County Council but answers to an independent Steering Committee.

The flexibility afforded by voluntary membership of the Forum allows local participation on a more informal basis. This clearly demonstrates fact that **voluntary** signing up to the principles of coastal zone management and participation in the development of policies within a plan will lead to active support for implementation. Seven years after its launch, the Solent Forum is still a strong organisation with widespread support for its work, which is mainly undertaken as Flagship Projects described below.

Coverage of the coast by fora is not universal. Indeed in some areas it is not necessarily appropriate or desired at the local level. However, in the absence of a national framework, the formation of such fora relies on the initiative and perseverance of one or two committed individuals. Such work has proved to be invaluable in improving co-operation and communication between various stakeholders and provides a mechanism for developing a co-ordinated approach to informing and influencing the statutory process. However, it is important to recognise the weaknesses of the voluntary approach, as at the very strength of its non-statutory nature can be its downfall as the policies produced may not have sufficient strength within the statutory framework. The harsh truth is, regardless of their success, voluntary initiatives are frequently the first to fall with increasing financial cutbacks.

3.3 Implementation of Policies

It is commonly quoted that non-statutory initiatives feed into statutory plans. Unfortunately, this is only achieved on an ad hoc basis. Not only because there is no compulsion for the planning authority to do this, but also because statutory planning has less flexibility in terms of timescale for amendment and review. A prime example of this is Shoreline Management Plans (SMPs). The non-statutory SMP process was invaluable in bringing together planners, engineers, ecologists and archaeologists in the context of coastal defence and it is fair to say that as a result relationships between these different disciplines have been strengthened. There is a greater mutual understanding of problems and in addition the SMP process has encouraged all parties to think more strategically about coastal defence issues.

The SMP has no statutory footing and, therefore, decisions about implementing the preferred policy options, developed after discussions and varying degrees of consultation, are not necessarily taken into account when making decisions on planning applications on the coast. An example of this is a coast protection scheme at Castlehaven, Isle of Wight, where English Nature who were members of the Steering Group have subsequently objected to the fulfilment of a "Hold the Line" policy which they did not object to during consultation. Furthermore, the concept of "no action" or "managed retreat" may not always be politically acceptable even if this is the preferred policy option based upon scientific evidence. There needs to be a form of 'safety net' in order to enable non-statutory policies such as these to be implemented.

The formation of voluntary groups and fora as a mechanism for improving communication and co-operation has been very beneficial, especially if the group can start the process with the production of a plan or strategy as a basis for discussion. However, as has often proved to be the case with voluntary plans there is a tendency that once completed, they are not implemented or updated; this is generally because there are insufficient resources available for the implementation phase rather than any lack of goodwill. The employment of dedicated project officers to implement voluntary ICZM policies and initiatives through mechanisms such as 'flagship projects' have proven vital to maintaining momentum and support.

Clear evidence of this is provided by comparing progress on the implementation of the two estuary management plans (EMPs) for the Western Yar and Medina on the Isle of Wight. Both plans were developed with extensive participation and consultation and there is wide support for the policies within them. However, when the Medina EMP was launched in May 1997 there was no project officer to take it forward and the process stagnated. It is highly likely that the Western Yar EMP (launched in December 1998) would have suffered the same fate had it not been for the appointment of a locally-funded Estuaries Officer early in 1999. Significant progress is now being made on the implementation of both plans for the benefit of the estuary users and interests.

In the case of the Solent Forum; their 'Strategic Guidance for the Solent' is being implemented through "Flagship Projects". These initiatives aim to improve the level of knowledge of the Solent and to continue to improve communication and dissemination to the wider coastal community. Events such as the Solent Science Conference, which brought together scientists and managers to produce a research agenda for the Solent; the identification of environmental indicators to assess the impact of plans and policies; and the wider dissemination of coastal management practice and initiatives through leaflets, publications and the Internet have all contributed significantly to the evolution and support for ICZM in the Solent. Information about these projects can be found on the Solent Forum's website: http://www.solentforum.hants.org.uk.

These examples show how the non-statutory approach can work to allow practitioners at the local level to tailor their requirements to specific local needs. Active, voluntary participation in such projects by a wide-range of users has fostered a feeling of ownership of plans and initiatives. The outcome is a positive attitude towards the principles of ICZM and a real desire to see mechanisms succeed in the longer term.

3.4 Use of Non-Statutory Structures to Fulfil Statutory Obligations

Through its work the Solent Forum is a well respected group, which has national recognition. The work of the Forum, like ICZM in general, is continually evolving and it is now being asked to undertake functions outside its original remit. For example, much of the Solent has been designated as either marine Special Area of Conservation (mSAC) or Special Protection Area (SPA) within the European Natura 2000 network. Under the Habitats Regulations (The Conservation (Natural Habitats, & c.) Regulations 1994) for which there is a statutory obligation to produce a single scheme of management, management groups and committees are to be appointed to take this process forward.

The vast majority of authorities required to manage the Solent mSAC are members of the Forum and its members have requested the Forum to be the mechanism for creating these groups. The Forum itself will not manage the SAC but will facilitate the process by which the SAC is to be managed. Although discussions on the technicalities are still underway, it has been agreed by members that meetings to develop the single scheme of management will follow each Forum meeting thereby reducing to a minimum the additional officer time, travel and subsistence costs required to support this legislation. There has been no indication of additional government funding to assist this process, which involves marine sites mainly outside local authority jurisdiction.

The role of the Forum has been vital in enabling this process as it is seen by members to be non-biased vehicle that allows all members views to be taken on board. This has alleviated many reservations about the process to develop the single scheme of management as concerns were repeatedly raised that larger organisations would 'steamroller' the process to achieve their own agendas.

3.5 Funding the Non-Statutory Approach

SMPs have now been produced for the coastline of England and Wales and with the assistance of grant aid from MAFF they have acquired a legitimacy. Similarly the estuaries initiative by English Nature provided pump-priming of estuary management plans. However, no such funding has been forthcoming for holistic management along the open coast. CZMPs have therefore relied too much on the willingness of statutory organisations to participate and to provide resources. Therefore, there are far fewer areas of open coast that are benefiting from integrated management on the ground as, in the absence of an ICZM plan everyday decisions are still taken on a sectoral basis.

ICZM, like any other work along the coast, can only occur if there are sufficient resources to carryout the work and implement ideas and projects. The same statutory organisations (e.g. local authorities, Environment Agency, English Nature and the local ports and harbour authorities) are asked continually for funding. With the reduction in local authority resources it is difficult to see how these organisations can sustain the funding for ICZM without any additional financial support from Government.

4. DOES ICZM NEED A EUROPEAN DIRECTIVE?

The paper so far highlights the fact that local, voluntary initiatives in the UK are currently working, but within a national policy vacuum. The national policy statements which the DoE claimed would inform local voluntary processes (DoE, 1993) have still not appeared as we approach the next millennium. There is a lot of activity at European level, which is now focussed through the Demonstration Programme. The question now has to be asked: should the EC fill this policy gap?

By looking at the work the EC has already undertaken it is obvious that the Commission has had a positive influence on the management of our coastal zones, for example, the Urban Waste Water Treatment Directive, the Bathing Waters Directive and the Environmental Assessment Directive to name but a few.

The EC now seems to be less inclined to support a legal instrument such as a Directive. This seems to be justified as coastal zone management is a **process** that varies enormously depending on the cultural values and economic pressures of each member state. There are major differences within the EU about social and cultural interpretations of rules and regulations and it is recognised that whatever mechanism is ultimately chosen by the European Union it *"must be sensitive to the legal variations between those States, and must permit the most suitable approaches adopted by each."* (p.23, European Commission 1999a). Despite institutional problems such as issues raised through the implementation of the Habitats Directive, our coastal zones and the natural environment are now much better protected than without European involvement.

The benefit of a Directive is that it is legally binding on member States and therefore can be enforced, thus overcoming one of the frequently expressed criticisms of the non-statutory approach. However, the voluntary process provides the opportunity to recognise the different cultural values in each of the member states. It has been shown that mechanisms for coastal zone management need to recognise the views of stakeholders and allow for public participation in decision-making and voluntary management agreements. It is felt that non-statutory advice backed up by financial incentives to follow this advice could be far more effective than prescriptive legislation. Not only is this likely to be passed with greater expediency by the European Parliament, but this more flexible mechanism would also allow member states to adapt the advice to their own particular legal, cultural, economic, social and environmental situations supported by an already heavily-legislated environmental protection framework.

5. CONCLUSIONS

In summary, the coastal zone is an extremely dynamic and complex environment which is extremely susceptible to change. The current statutory process in the UK is very long-winded, taking approximately two to three years to put in place statutory development plans that will then stand for another 10 to 15 years allowing for little flexibility. The voluntary approach is seen as laudable in that it promotes consensus, is adaptable and able to respond to changes in circumstances. However, it has been argued that such a policy ignores the fact that there is a need for some clarification of responsibilities, which would benefit public perceptions and ease the path of consultation and participation.

A European Directive would require member states to bring about legislation which, if too prescriptive could hinder the process of ICZM. Legislation in itself will not bring about improved coastal zone management, there needs to be a political will to take the process forward. Responsibility ultimately lies with the users of the coastal zone and their willingness to adhere to any framework. As it currently stands in the UK the ICZM process relies on the willingness of interested parties to contribute time and resources, there is no statutory back-up and due to budget constraints facing many participating organisations, non-statutory initiatives are always under pressure and relying upon short-term funding for their survival.

In response to the question of the advantages of the voluntary approach, the impact has been high, a great deal has been achieved over past decade in gaining acceptance of the need for a holistic, multi-sectoral process. There is a much better recognition of the wide range of interests and the need to take all of these into account when trying to resolve issues. Public participation and consultation has enabled huge unpaid resources to be tapped by developing a sense of ownership in plan development and implementation. There is clear cross-linking between the statutory and non-statutory plans in many cases, both vertically and horizontally.

The disadvantages of the voluntary process in the longer term are that the goodwill put into developing these initiatives often suffers due to their non-statutory status and thereby inability to enforce the policies developed within them. It could be argued that making them statutory would derive them of their flexible approach, however, without a national framework, rather than a cascade, what we have is a fountain of plans which is at risk of looking decorative, with a high feel-good factor but which ultimately brings us no closer to the fundamental goal of sustainable use. The authors believe the voluntary approach can work, but will struggle in the longer term without an enabling framework incorporating clear support from both the EC and national government.

REFERENCES AND BIBLIOGRAPHY

Arun District Council (1993) *Local Coastal Management Plan* (Arun District Council).
Belfiore S.1996, *The role of the European Community in the Mediterranean coastal zone management, Ocean & Coastal Management*, Vol. 31, Nos. 2-3, pp. 219-258, Elsevier Science Ltd. Northern Ireland.
Chichester District Council (1995) *Coastal Zone Management Plan* (Chichester District Council, Chichester).
DoE (1992) Planning Policy Guidance 20: Coastal Planning (HMSO, London).
DoE & Welsh Office (1993) *Managing the Coast A Review of Coastal Management Plans in England and Wales and the Powers Supporting Them* (Crown Copyright).
DoE (1993) *Coastal Planning and Management: A Review* (HMSO, London).
DoE (1995) *Policy Guidelines for the Coast* (Crown Copyright).
DoE (1996) *Coastal Zone Management Towards Best Practice* (Crown Copyright).
European Commission (1999a) *Towards a European Integrated Coastal Zone Management (ICZM) Strategy General Principles and Policy Options* (European Communities, Luxembourg).
European Commission (1999b) *Lessons from the European Commission's Demonstration Programme on Integrated Coastal Zone Management (ICZM)* (European Communities, Luxembourg).
Hampshire County Council (1991) *A Strategy for Hampshire's Coast* (Hampshire County Council, Winchester).
Huggett D. (1996) *Progressing coastal zone management in Europe: a case for continental coastal zone planning and management.* In eds. Taussik J & Mitchell J (1996) *Partnership in Coastal Zone Management* (Samara Publishing Ltd., Cardigan) pp.47-56.
McInnes R M (1994) *A Management Strategy for the Coastal Zone* (South Wight Borough Council, Ventnor).
New Forest District Council (1997) *Coastal Management Plan* (New Forest District Council, Lymington).
SNH et al, (1997) *European Marine Sites, an introduction to management* (SNH, Perth).
Solent Forum (1997) *Strategic Guidance for the Solent* (Hampshire County Council, Winchester).
West Sussex County Council (1994) A Coastal Strategy for West Sussex (WSCC, Chichester).
Western Yar Steering Group (1998) Western Yar Estuary Management Plan (Isle of Wight Council, Newport).

Tourism and resort action plans – identifying a methodology

Peter Lane, Acting Director Community Leisure, Libraries & Tourism, Redcar & Cleveland Borough Council, U.K.

Abstract

The Coastal Zone in general and seaside resorts in particular provide a major environmental and economic resource. Many of the resorts created at the turn of the century are in decline. Newer resorts created in the 1960s are failing and the fragile coastal environment is threatened by development pressures. There appears to be a lack of understanding about the value and importance of resort and coastal areas and an inability to address fundamental problems. Such difficulties are not unique to any particular country as it is apparent that the underlying problems of coastal resort areas are widespread throughout Europe. With varying degrees of success attempts are being made to redress the balance and effect a process of regeneration. This paper covers some of the resort regeneration issues and outlines the work that has been carried out by two European networks in establishing positive and cost effective guidelines for action.

1. Context

1.1 For over 200 years seaside resorts dominated the tourism scene. Around 20 years ago they accommodated some 75% of holiday makers and now they account for around a quarter of tourist trips and a third of staying visitors expenditure. In the 1950s though, the traditional seaside holiday market began to change significantly. Due to complacency or a lack of understanding changing trends were not identified or responded to. The apparent survival of resorts into the 1970s, due partly to overall market growth, masked the truth that the seeds for decline had already been sown and were beginning to take root.

1.2 The main reasons for decline are well known and documented and include:-

- the increase in car ownership leading to improved personal mobility and a broadening of choice;
- competition from newer destinations offering better value, modern facilities, more reliable weather and a travel experience;
- competition from inland centres and areas, self-catering holidays and the new wave of holiday centres;
- a lack of investment;
- a switch in holiday trends and growth of the day visitor market.
- seasonality.
- a lack of identity and focus.

- complacency and a failure to respond to changing market trends.
- environmental degradation.
- the insensitivity of resort architecture particularly that produced during the 1960s.

1.3 The affects of change and decline are not confined to traditional resorts. Those built in the 1960s, particularly catering for the mass market in the Mediterranean, are increasingly affected by changes in market trends and the need to revitalise an outdated product. Image and quality are extremely important factors.

1.4 Nevertheless pressure for new development in key locations continues. Such pressures have a significant consequence for the coastal environment. The sprawl of tourism threatens fragile areas and can destroy the very essence of the originally marketed destination through over use, congestion and insensitive development.

2. **Responses To The Problems**

2.1 Up until the mid 1970s competition and rivalry between resorts for the reducing amount of trade was sufficient to prevent them from admitting that problems did exist. By then it was becoming self evident that such problems were widespread and deeply rooted and that fundamental action was needed. This realisation was paralleled by an increasing awareness of the significance of tourism as an important economic sector and a force for change.

2.2 In European terms it is the second largest industry and is one that unlike any other sector, is estimated to continue growing significantly. Yet the inadequacy of local tourism statistics makes it difficult to prove that there is some positive economic value in sustaining and developing the industry.

2.3 In 1991 the English Tourist Board recognised the plight of resort areas and produced a report on their likely future. The report provided a comprehensive overview of resort issues and concluded that there was a need for:-

- An understanding of the visitor market.
- A coherent strategy.
- Tourism to be considered as part of an area economic strategy.
- A corporate approach to tourism by all council departments.
- The creation of a vibrant image with a broad based appeal.
- Partnership with local business and residents.
- Improved marketing.

2.4 Their criteria for creating sustainable resort regeneration were:-

- The concentration on regeneration of tightly defined core visitor areas.
- The encouragement of mixed development.
- The accommodation of residents' needs.
- Maintenance of a balance between staying and day visitors.

The following should be added to the list:-

- The creation of an identifiable and marketable image.
- The development of partnership with the private sector and with resident, business and community groups.
- The raising of standards of visitor care through training and targeted investment.

2.5 A committee of inquiry into resort areas, established in 1993 by the Association of District Councils, also concluded, in their document *Making the Most Of The Coast,* that:-

- Seaside resorts do have a future in tourism but the nature and scale of the industry will vary between resorts.
- Partnership arrangements are essential.
- Local Authorities need to be committed to tourism.
- Coastal areas across Europe share a pressing need for redevelopment and investment.
- Service standards must be improved.
- The quality of the built and natural environment must be maintained to high standards.

2.6 More recently the British Resorts Association, in their report "UK Seaside Resorts - Behind the Facade" have identified the special and unique problems of such areas and the need fro greater recognition and support.

2.7 On a broader scale other areas were beginning to look at resort issues. The 'Fresh Noses' scheme along the Dutch North sea coast recognised the need to 'freshen' the image of their resorts. In Mallorca stringent environmental controls and a rationalisation of accommodation were instituted in the 1980s to alter the islands image and change its market characteristics.

2.8 Individual resort areas were also, independently, trying to come to terms with their declining trade and loss of market share by looking more closely at their assets and potential markets.

2.9 In general though the problems in Europe were seen to affect mainly the traditional resort areas and those created hastily in the boom period of the 1960s. Resolving their problems was thought to be almost impossible without the injection of substantial finance.

3. Establishing A Way Forward

3.1 Two European Network Projects have looked specifically at the problems of resorts and at the potential for achieving effective revitalisation with minimum funding. Whilst these projects have concentrated mainly on seaside resorts it is considered that the conclusions drawn and principles suggested are applicable to the whole range of potential visitor destinations.

3.2 The first of the European network projects was based around the following eight primary aims:-

1. To appraise the relevance of the following initiatives in securing economic regeneration through sustainable tourism developments and assess their role as part of Regional Development Policy.

   a) The development of major new leisure projects capitalising on the areas assets and, where eligible, using EC funding.
   b) The interpretation of an area's history and, by using historic buildings as the basis, the creation of innovative visitor attractions.
   c) The extension of the holiday season in time and space including the use of events to stimulate local communities, create confidence and act as a focus for attracting visitors.
   d) The revitalisation of historic and/or unique settlements.

2. To assess the importance of environmental issues in resorts and their incorporation into new development initiatives.
3. To determine the relevance of marketing and promotional activities, the way they are carried out, the levels of need and who should carry them out.
4. To assess the value of partnerships and other associations.
5. To appraise and redefine the role of the seaside resort as a visitor destination.
6. To define other possible approaches and by study and refinement produce development guidelines for other authorities throughout Europe.
7. To act as a core group to initiate and try out new policies as a guide to national and European decision making.
8. To promote long term working relationships between the project partners.

3.3 Eight partners were involved in the project and each had assumed that their coastal/resort problems were unique. It soon became apparent that similar problems existed in all the resort areas, in varying degrees of intensity. Interestingly even the myth that cold (northern) water resorts are fundamentally different to those in warm water areas was dispelled.

3.4 The most significant overall conclusions of the work were that:-

1. There was a need for clear and comprehensive planning for resort areas.
2. The economic significance of tourism needs to be more clearly recognised.
3. The community needs to be involved in the planning process.
4. Resorts need to be seen as part of the wider geographic region.

3.5 Other key findings were that:-

- There is a need for resort regeneration.
- Tourism is a prime function of resort but there is a general lack of understanding of the economic role and a failure to look comprehensively at the issues involved.
- Economic activity is centred around small to medium sized enterprise which are vulnerable to change and lack the stature to secure adequate investment or reinvestment funds.
- Existing resort areas provide a tremendous tourist resource and can handle, with minimum impact substantial numbers of visitors.
- Resorts need to define and develop their own unique qualities to complement rather than compete with other areas.
- Resort issues need to be tackled comprehensively within the individual resort and on a wider geographical basis.
- There is a need for clear strategic planning at the local, regional, national and European level.
- Success depends on the level of commitment from all the parties involved in investment of the host community.
- Pump priming finance is an essential ingredient of success.
- The problem of resort regeneration and development is an issue throughout Europe for which specifically targeted structural funding is required.
- There is a need for a European forum to consider coastal and resort issues, promote new initiatives and monitor and evaluate trends.

3.6 The second project developed from the principles established in the first and was based around the need to demonstrate and evaluate regeneration techniques for developed coastal areas and refine them into a transferable model for use in different types of resort areas throughout Europe.

3.7 Much of the work was based around the formulation and testing of guidelines for the preparation of Resort Action plans for each of the areas involved in the project. Such plans were backed up by the implementation of a variety of demonstration actions designed to illustrate commitment to the revitalisation process.

3.8 The main conclusions of the project are that:-

- Resorts are a key function and element of the coastal zone and need to have prominence in planning for it because:

    a) they have capacity to handle with minimum impact significant number;
    b) there is continuing pressure for continuing development in sensitive coastal zones due to the apparent lack of available development land;
    c) resort visitors have an impact on a wide area around it;
    d) there is a need to manage visitor usage;

- Image and identity is a critical factor in placing a resort in the market.

- The quality of the visitor services and visitor experience is a key determinant of a resorts long term future success.

- The environmental impact of tourism needs to be carefully addressed not just in physical terms but on the impact on infrastructure and energy usage.

- Comparative indicators need to be produced to:

    a) evaluate the value of tourism in a consistent manner;
    b) be able to compare qualitatively individual resorts;
    c) identify the nature of individual resorts.

- The need to involve the community in the decision making process.

- Comprehensive resort strategies are an essential tool for the effective planning of coastal areas and can be produced relatively simply by using the common format devised as part of the project.

- Regular monitoring of the visitor market is essential to identify new opportunities and specialist market initiatives.

- Resort revitalisation should be given greater support by national and European agencies.

- Pump priming initiatives can help to achieve confidence in an areas future and generate a commitment to revitalisation from local organisations.

- The principles of resort revitalisation can be applied to any form of visitor destination.

4. Developing A Methodology

4.1 An assessment of relative problems and the drawing of conclusions from a comparative analysis of different resort situations was an important part of the second project. Of more practical benefit has been the preparation of a methodology for producing Resort Action plans and testing them in different situations. Through this process it has been possible to produce a transferable model that, it is felt, can be used in all the member states.

4.2 It is hoped that by adopting the model those managing resort areas will be equipped with an organisational and procedural framework that will enable them to produce and implement comprehensive regeneration strategies in a cost effective manner. This should allow them to make the most of their visitor resources in a sensitive and sustainable manner.

5. The Tourism and Resort Action Plan Process

5.1 Extremely detailed, complex plans based n the collection of a mass of data have their place but generally resort tourist areas need a practical easy to understand formula that involves local people and helps guide action and development over the short to medium term. Regeneration needs commitment from a variety of partners and, therefore, the plan process needs to be capable of being assimilated easily by people with a variety of particular interests. The plans produced as part of the process need to be action orientated, acceptable and realistic.

5.2 It also has to be remembered that a plan is only part of the process of revitalising a resort. However it is an important first step in that it provides evidence of a logical process of analysis and plan formulation which can be used to prove the case for action. The actual process of plan preparation is often as important as the resulting document as it creates a framework for discussion and debate and helps to focus attention on key or emerging issues. Debate and discussion necessitates involvement and it is a prerequisite of plan preparation that interested organisations and individuals are involved at the earliest possible stage. This helps to create commitment and ownership and builds confidence within the community.

The plan process also provides a discipline against which performance can be monitored. This reinforces the need for such pans to be realistic, achievable and, more particularly, acceptable to those who are likely to be affected or influenced by them.

5.3 The process devised has been separated out into eight distinct stages as follows:-

STAGE 1	Realisation/ Creating A Vision	• Realising that action needs to be taken to deal with specific problems. • Establish dialogue with the host community. • Looking at the current resort vision. • Adjusting the current vision or creating a new one to relate to new and emerging circumstances.
STAGE 2	Political Ownership & Commitment	• Stimulate political debate. • Establish a political consensus. • Secure political commitment.
STAGE 3	Creating a better understanding (Date Collection)	• Carry out visitor surveys. • Establish database. • Carry out environmental audit. • Analyse existing plans. • Establish process for obtaining community and service providers views and to involve them in the process.
STAGE 4	Ideas generation & Data Interpretation	• Carry out SWOT Analysis. • Secure ideas from community groups and organisations. • Feed data analysis into the consultation process.
STAGE 5	Identify the key issues.	• Establish the key issues, list and seek agreement from consultation groups.
STAGE 6	Policy Formulation	• Relate key issues to policy options. • Test policy options against broader issues e.g., likely finance available, potential partnership opportunities, community needs, sub regional/regional issues.
STAGE 7	Specific Actions	• Definite detailed actions, identify body responsible for implementation. • Set target date deadlines. • Obtain full agreement to the plan content.
STAGE 8	Plan Monitoring	• Establish monitoring timetable and determine data to be collected for monitoring purposes. • Continue to collect and analyse the relevant data on at least an annual basis. • Carry out annual visitor surveys. • Establish sources of non visitor information.

5.4 Interestingly the process outlined parallels the framework suggested for the preparation of Local Agenda 21 Action plans. It is possible, therefore, from the outset to decide whether or not to pursue the Tourism or Resort Action Plan as one that fully embraces the principles of sustainability.

6. Relationship to Coastal Zone Management

6.1 Although resorts are a dominant feature of coastal zones and are centres of intense activity, rarely do they figure significantly in overall coastal zone planning. It seems more appropriate, in comprehensive planning terms, to ensure that all elements of the coastal zone both developed and undeveloped are dealt with. Only through effective integration will it be possible to pursue visitor management plans and more positive moves towards the sustainable tradition of the coastal zone.

6.2 The resorts produced as part of the Resort Regeneration process echo this fact. It would help, therefore, if a better understanding could be reached between those involved in the management processes for the developed and undeveloped coast. Groups, such as the Coastal Special Interest Group, need to look at the Coastal Zone in a more comprehensive way and include those involved in managing existing resort areas more effectively in the process.

Flood and Coastal Defence Aims, Objectives and Targets

R G PURNELL, Chief Engineer and B D RICHARDSON, Regional Co-ordinator,
Flood and Coastal Defence with Emergencies Division
Ministry of Agriculture, Fisheries and Food, London, England

INTRODUCTION

The last ten years have seen significant improvements in the way that flood and coastal defences are delivered in this country. Understanding of coastal processes has improved with the potential for a more scientific approach to the design of coastal defences. In conjunction with the technical improvements policy and approach by central and local government has evolved. In 1993 flood and coastal defence policy was re-examined and re-stated in the Ministry's Strategy for Flood and Coastal Defence in England and Wales. This gave the published aim of the MAFF and Welsh Office policy as "to reduce the risk to people and the developed and natural environment from flood and coastal erosion by encouraging the provision of technically, environmentally and economically sound and sustainable defence measures". A clear and transparent aim for the authorities and other contributors to deliver. This was backed up by improved guidance and encouragement for a strategic approach to the delivery of flood and coastal defences.

Time has moved on and it is now appropriate to review those achievements and seek to further improve delivery of this important public service.

THE PROBLEM

Successive Governments have long recognised the potential impact of flooding on the economic wealth of the nation and as a result invested significant sums of taxpayers' money in order to reduce and mitigate that risk with the express purpose of protecting the nation's economic wealth, health and prosperity. In the past the need for works to alleviate flooding or coastal erosion was only recognised after a damaging event. The result of this approach tended to be a low level of investment leading up to a damaging event after which investment continued at a much higher level until the impact of the flood was forgotten. The approach has progressed significantly and operating authorities are now encouraged to analyse the flood and coastal defence risks over longer timeframes with the intention of reducing the risk of economic loss before the inevitable event occurs. As this approach is more widely adopted, major damaging floods should become more rare, although the impacts when they occur may be more significant. The temptation remains to reduce investment in the short term. Where this occurs the potential residual damage will increase and be confirmed by the inevitable event. If this intermittent approach to investment is to be avoided then a closer correlation is necessary between the level of investment and the real size of the problem to be addressed.

One objective of long term strategic planning and shoreline management plans was to provide a clearer picture of the overall extent of the flooding and coastal defence problem in England.

Coastal Management: Integrating science, engineering and management, Thomas Telford, London, 2000

Where well developed strategies have been agreed there is a useful basis for long term investment planning but it could be some time before coverage is universal.

A change of administration in 1997 and the introduction of the Comprehensive Spending Review (CSR) resulted in a requirement for close programme justification. Every Department was required to review the need and effectiveness of their investment programmes. The exercise was influential. Although a broad brush approach was adopted the study noted that over 2,200 sq km of land in England is at risk of tidal flooding and, of this, some 10% is classified as urban. In addition some 1,000km of the coastline is at risk of erosion with 90% of frontages currently protected. The potential average annual damage in these coastal areas without the intervention of man approached £1bn. A significant sum although investment over many years has reduced this to more acceptable levels. The question remained, however, as to whether further investment to reduce the residual damage any further was worthwhile and, if so, what the optimum level of investment should be.

The CSR study clearly demonstrated a need to gain a better understanding of national flood and coastal defence risk. For this reason MAFF has now commissioned research which will attempt to better define the current magnitude of the potential national economic loss. If understanding of the problem is to be improved it will be necessary to monitor and review changes over future years in terms of natural change (including climate change) as well as asset conditions and other man made impacts such as development and demographic change.

CURRENT LEVEL OF INVESTMENT AND PRIORITISATION

Total expenditure on flood and coastal defence is in the region of £400m per annum with probably half being spent on the coastline and the remainder on rivers and other watercourses. Of this total, about £140m per year is spent on capital improvement works of which some 60% is spent on the coastline. Currently decisions on expenditure are, in the main, taken locally although the influence of availability of central government funding and revenue support grant cannot be ignored. Local representatives consider their flood and coastal defence problems and on the basis of the resulting investment strategy bid against the available national funds. Inevitably, perhaps, the call on national funds has always exceeded supply, a situation which had, in recent years, become more pronounced. On the basis of available long term plans this did not seen likely to improve.

In an effort to optimise the allocation of available national funds a prioritisation system was introduced in 1997. The overall priority score is based on priority assessment in three distinct areas. Points are scored on the principal land use of the area to be protected, with urban areas scoring higher than rural areas; on urgency, essentially a measure of whether the scheme could be delayed; on the authority's initial assessment of the benefits and costs of the proposal, with the highest score going to these schemes with the highest rate of return. Funds are restricted to those schemes scoring highest with the threshold score being set annually based on a balance between demand and available funds.

We believe that the system has worked well with priority clearly being given to those schemes likely to provide the greatest protection to life and property, and which cannot be delayed. However with the priority score threshold now at 23 there is a question as to whether the available funds match the totality of the nation's need for flood and coastal defence. The original presumption was that the priority score system would delay only those schemes which lie at the margins of acceptability. With a priority score at 23 (out of a potential maximum of 30) clearly some urgent high priority schemes are being delayed.

This evidence from the priority score system tends to reinforce the conclusion of the CSR that the total national level of expenditure on flood and coastal defence is insufficient to meet demands. However, the evidence is far from conclusive with each review suggesting pressure of demand from a different area. It is reasonable therefore to more fully consider the potential impact of flooding and erosion on the national economy to give greater certainty to the derivation of reasonable levels of investment and priority. This will enable more robust bids to be made in competition with other government spending priorities. It will also give us an indication of whether changes are required in the priority system in order to more appropriately optimise the allocation of funds.

Looking to the future we must put ourselves in a position where the impact of investment decisions (increases or decreases) can be considered and the reasonable future requirements for funds more easily judged. Making best use of taxpayers' money calls for decisions on investment; waiting until floods provide evidence of under investment is not an efficient economic option.

TARGETS AND ACHIEVEMENTS
Clearly the proposals so far outlined in this paper will better enable us to understand the pressures on flood defence and coastal erosion and decide on the investment levels required. However, whilst we have a clear aim and objectives, there is no reliable and consistent means of measuring delivery. Is the Government's flood and coastal defence aim being achieved? The Ministry has decided therefore to identify a series of targets that are necessary to facilitate a more certain delivery of national policies and objectives for flood and coastal defence. In particular this will initially be achieved by a more systematic gathering of information about the nature and status of defences and by having in place arrangements to assess the integrity of defences or facilitate any remedial action that may be necessary.

The targets so far identified in the interim proposals are a first step in linking performance to the MAFF objectives and the underlying Government policy aim for flood and coastal defence.

Further consultation is being undertaken on future targets which will measure delivery of the principle MAFF objectives. To achieve the objective of encouraging the use of adequate and cost effective flood warning systems, the need for a target for the provision of flood warning in terms of population and area covered has been agreed with the Environment Agency. A necessary complement to this will be targets to ensure that appropriate mitigation and emergency response actions will be delivered.

Another objective is to encourage the provision of adequate, economically, technically and environmentally sound and sustainable flood and coastal defence measures. Identification of assets which provide a flood and coastal defence service either owned or operated by the Environment Agency or others is a first step to measure achievement of this objective. This will clearly involve a large amount of information and urgent consideration is being given to the structure of a database which will make best use of this information. It is likely that this will need to clearly link all defence infrastructure with the areas and assets at risk to which protection is provided. A useful starting point should be the flood risk maps currently being produced by the Environment Agency and corresponding coast erosion mapping. Following on from the collection and analysis of such data it is only a short step for authorities to provide a prioritised forward plan of capital and maintenance work in order to ensure that those flood and coastal defence assets are, where required, maintained to an appropriate level when compared with the risks involved. From such information we should then more clearly see whether sufficient funds are being invested to maintain or reduce the overall national flood and coastal defence risks.

We have previously made clear that the programme of improvement and replacement works should be tied to a long term strategic plan and it is likely that the production of strategy plans and shoreline management plans will be taken into the system of targets. There are also likely to be targets to measure achievement of international obligations with regard to environmental assets.

THE FUTURE
What we have outlined so far marks a significant step forward in the planning of flood and coastal defence in this country. For the first time we should have a much clearer indication of the size of the national risk. This will inform decisions on the level of investment needed to protect the nation's economic wealth and the consequences of not meeting that level of investment. Such information should reinforce long term investment plans and help to avoid short term changes in provision. It will also provide a comprehensive list of the flood and coastal defence assets and their condition; however we still need to go further. It should be possible, indeed it is desirable, to fully integrate problem definition, measurement of achievements and targets. The setting of achievable targets depends on an understanding of the nature and scale of the problem, which will clearly reduce as targets are achieved. An important aspect of this will be an information system which is based on both flood and erosion risk areas. This should provide a clear connection between the problem and the condition of the asset; and allow any changes in the size of the problem with time to be assessed. It should then be possible to align targets more closely with economic and social impact. More important, perhaps, such information should be made available in the public arena so that individuals and local groups are in a better position to make more informed decisions themselves.

CONCLUSIONS
The move towards the production of high level targets for flood and coastal defence seems at first sight a small and straightforward step. However, its significance in terms of forward planning and achievement of the Government's flood and coastal defence policy aim cannot be overstated. Adoption of such measures will be of significant benefit to all involved in flood and coastal defence. In particularly the operating authorities will have a much clearer framework in which to operate and achievements should receive greater recognition. However it has a downside; absence of achievements will also be much more publicly visible.

REFERENCES
Strategy for Flood and Coastal Defences in England and Wales. MAFF/Welsh Office 1993. MAFF Publications PB1471

Coast Protection Survey of England, Summary Survey Report
MAFF (1994). MAFF Publications PB1667

NRA Sea Defence Survey, Phases 1, 2 and 3. NRA (1992)

MAFF Funding for Flood Defence and Coast Protection:
Notes for Guidance MAFF (1997)

Comprehensive Spending Review, Working Group on Flood and Coastal Defence: The Risk of Flooding and Coastal Erosion in England and Levels of Resources Needed (Unpublished). MAFF (1997) Summarised in the Government Memorandum of Evidence to the Agriculture Committee, Sixth Report (Session 1997/98) Flood and Coastal Defence, Volume II (ISBN: 0-10 - 554677-1)

Interim High Level Targets for Flood and Coastal Defence and Elaboration of the Environment Agency's Flood Defence Supervisory Duty. MAFF (1999)

Consultation Paper on High Level Targets for Flood and Coastal Defence. MAFF (1999)

Discussion

COMPARATIVE COASTAL MANAGEMENT

G. Guthrie, *Paper 15*
My question is to John Pos, regarding creating opportunities for land use in the environment. Planning is clearly the starting point for developing the strategy that you described. The emphasis in the UK tends to focus more on where the coast wants to be and the land use plans tend to be fitted in around this. Do you feel that in the UK there is scope and need for stronger planning of what society wants or needs in terms of land use and environmental resources, and that those requirements should form our starting point for actually managing and engineering coastal processes? Also, are our current procedures possibly precluding that attitude?

Dr J.D. Pos, *Paper 2*
I think the Dubai situation differs greatly from the UK situation in that we have much more scope to be planning-driven since the consultation process is far less onerous. In fact, most of the consultation is conducted with the Municipality who in turn consult with the Ruler. In the UK you have great difficulties in actually pushing through grand schemes like this because of the many tiers of consultation that you have to go through before anything can be achieved.

P. Sayers, *Paper 11*
A question for Professor Hettiarachchi: in the UK, telephone consultation or even Internet consultation can be effective in addition to face-to-face surveys; what particular difficulties did you have in disseminating your concepts of coastal zone management to the general public?

Dr S.S.L. Hettiarachchi, *Paper 1*
With respect to the consultation process, as I mentioned there have been a lot of protests with regard to coastal development activities. For example, when revetments are constructed, certain sections of the fishing community are affected because they no longer have places to moor their vessels, etc. The Coast Conservation Department tends to handle these issues through local councils and through the environmental impact assessment procedure. The EIA procedure ensures that the people affected by any scheme are consulted and their opinion taken into account prior to development work. In fact, many major projects have not gone ahead because they have not been able to pass the EIA stage. The Government is currently having great difficulty in getting the approval for an EIA for a coal-powered plant which is high on the coastline, due to the level of protest from the local people, including religious organisations. We do have very strong NGOs in Sri Lanka and this seems to be the current trend in the developing world. There are conflicts as well, but very strong pressure can be brought to bear by the NGOs.

Coastal Management: Integrating science, engineering and management. Thomas Telford, London, 2000.

K. Millard, *Papers 10 and 11*
A question for Peter Barter: the process of formulating the Barbados CMP has resulted in an impressive data set of many coastal parameters. After the initial design phase, what mechanisms are foreseen in the CMP for the long-term custodianship of the data so that it can be effectively re-used in 10, 20 or 30 years' time?

P.W.J. Barter, *Paper 3*
The current arrangements in Barbados involve a long established coastal zone management unit, staffed by approximately 20 people from different categories. They are charged with implementing the plan and keeping up to date any data as it is received. The Coastal Zone Management Act also requires them to review the plans every five years. Obviously, part of that process will in fact be covered by this Phase II project, which is linked to investment that is due to take place in a couple of years' time and will probably last for around five years. However, I can see the demand for a greater custodianship arising and certainly there is sufficient strength of interest within the client base to ensure that it is properly managed in the future.

I. Heijne, *WS Atkins Consultants Ltd*
The four papers this morning cover the widely differing economic status of countries, from Sri Lanka as a developing country to the UAE. With all the plans proposed being very forward looking, have the authors any views on the lessons learned and how they may apply to the UK?

Delegate
The idea of having the planning component more strongly involved with the management and development of the coastline is an issue which should be addressed in the UK.

Dr S.S.L. Hettiarachchi, *Paper 1*
Just a brief comment; I accept the fact that Sri Lanka is a developing country and that we have limited funds. We have received generous assistance from many countries, but I believe that it is important that the funds are well utilised. In fact, there is a large CZM project funded by ADB which is due to start very soon, and the funding agency will review the whole Sri Lankan experience prior to the release of funds. Of course, it is difficult to undertake large-scale development work if we do not get external funding, which also has to be matched by the Sri Lankan government as well.

Y. Imamura, *Paper 4*
I would like to comment on the situation in Japan. As you know, Japan has been developing very rapidly since World War II and the total budget for public works, including coastal management, has been growing very rapidly. So we have the opportunity to look at ways of preventing disasters and saving lives and property. However, I think our economy has reached a turning point. We cannot expect further growth and, moreover, we are facing particular financial challenges. For example, the ratio of elderly people is growing rapidly so we have to spend much more on our welfare system, and we have also constructed a lot of infrastructure since World War II, so we expect that by 2010 to 2015 more than half of the budget of the public purse will be spent on maintenance not construction. Therefore, we have to look at the most efficient use of our investment for all public works, and coastal management is included that category.

P. Murby, *The Wildlife Trusts*

I would question the effectiveness of CVM to capture ecological values (most of which cannot be expressed in money terms). More importantly, the valuation process described appears to lack balance because intangible benefits are measured but intangible costs are neglected. The effect is likely to exaggerate the case for capital investment.

Dr D. Brook, *HOB Planning and the Physical Environment, DETR*

My first comment is that when I saw Fig. 1 on Dr Hettiarachchi's paper I was appalled at the specification of the coastal zone to the 300 m limit as being too limiting in view of the need to consider the zone in the context of the hinterland. The paper mentions the criticism of the CCA definition and that it is being revised. Is it reasonable to be so specific in defining a coastal zone? Or is it more reasonable to accept that the zone is only vaguely defined because of the wide influences that derive from the hinterland?

Dr S.S.L. Hettiarachchi, *Paper 1*

There has in fact been wide criticism on that issue, but it was necessary because of the very high level of degradation which was present at that time. This is now being reconsidered and will probably be changed in the future. I have a strong feeling that if we did not adopt this specification it would not have been possible to control the degradation since we had erosion right round the country, as shown on the map. If erosion had not been controlled most of our infrastructure would have been in danger. So that is the main reason for the narrow specification which could be well managed with considerable authority.

A. Wilks, *Chairman, Scottish Coastal Forum*

A question to Dr Hettiarachchi: the UK has adopted a non-statutory, voluntary partnership, approach to CZM and, as yet, there is no consensus on an agreed definition of the coastal zone: what made the Sri Lankan government realise that not only a definition but also a Coastal Zone Act was essential?

Dr S.S.L. Hettiarachchi, *Paper 1*

I believe that interest in coastal erosion came into focus in the late 1960s when several investigations were carried out with the assistance of overseas consultants. At that time, the Government had no option but to address those issues as a matter of urgency in view of the prevailing situation. Had they not, the consequences would have been very severe. So I believe that is why they had to adopt such rigid specifications for the coastal zone. Even after nearly two decades of coastal zone management I still believe such regulations are necessary because without such regulations it is very difficult to control. The Act is the legal arm for the implementation of the CZM Plan.

The setback and variance have been changed over the past decade. A more relaxed attitude is taken towards it since studies have been undertaken and the levels of erosion monitored.

C. Brown, *Seabee Developments*

Given the worldwide (albeit northern hemisphere only) spread of the presentations in the first four papers, and the topic of 'Comparative Coastal Management', I would be pleased to discuss a subject at the heart of the matter, yet barely touched on in the presentations — that of sea level rise. Could each of the authors comment on the consideration given in their plans, i.e. the method/model used, and how far ahead they look. What are the effects on their management plans of continued or accelerated rise after this period.

Dr J.D. Pos, *Paper 2*

Sea level rise is dealt with in Dubai in the same way as in the UK. In fact, most of the design methodology that we use for appraising the long term evolution of the coastline is, I am sure, very familiar to practically everyone here, so I do not think I can really add much to this.

P.J.W. Barter, *Paper 3*

As far as the Barbados situation is concerned, we certainly considered the existing guidance, but we do have the additional benefit in Barbados that the island is rising and therefore does not have quite the same problem as some other areas. But certainly allowances were made in all the analysis work that we did.

Y. Imamura, *Paper 4*

In Japan, the problems of both sea level rise and global warming generally receive less attention than in European countries or the UK. Therefore I must say that at the moment our coastal management programme does not consider sea level rise. However, although our financial situation is becoming very difficult, we still have access to the financial support and to the technology to cope with these issues when the problem becomes apparent. At present the situation does not impinge on the Japanese as it does in Europe where perhaps the results of global warming and sea level rise are more apparent, so these issues do not receive the same level of attention.

Dr S.S.L. Hettiarachchi, *Paper 1*

I am very pleased to say that the Sri Lankan Ministry of the Environment is currently preparing a national action plan for climate change, and sea level rise will come under that. In fact, I am in charge of the sea level rise section and we are coordinating with all the government departments who are stakeholders along the coastal zone so that they will also take into account the impacts of sea level rise. The main area is salt water intrusion because we have large areas of low lying agricultural land which will potentially be affected and resulting from that we are proposing an inventory of low lying areas, etc. Incidentally, one interesting point is that I think global concern over sea level rise was really focused in 1989 after the Maldives Conference, which was just after the first ICE Conference on Coastal Zone Management.

PREDICTION AND MANAGEMENT OF CHANGE

J. Andrews, Posford Duvivier

A question for Steve Hayman: we have been looking at similar problems with regard to shingle ridges as part of a strategy study for the Selsey Peninsular on the south coast. In that case we developed an option which involves recharging the landward face of the ridge. This has the advantage of allowing some regression to a less exposed line, it also enables the beach to take up a flatter and more stable profile. Had you considered that option for Salthouse? Also, continued working of the beach with bulldozers can be expected to degrade the beach material itself to a smaller particle size. A further result of this activity is likely to be lowering of the lower foreshore. Have either of these issues been identified as a problem with the present management policy at Salthouse?

Dr N. Pontee, *Paper 12*

Steve Hayman mentioned that one of the problems involved in reprofiling gravel beaches with bulldozers was that this led to a lack of sand matrix within the gravel. He suggested that this was a factor in reducing the flood defence capability of the beach. From my experience, with mixed sand/gravel beaches throughout the UK, the key to the stability of such features under wave attack is their permeability. The high permeability of gravel results in high rates of wave energy dissipation.

On many recharged or reprofiled mixed sand/gravel beaches, the presence of sand in the gravel, coupled with compaction of sediments due to the use of plant, are important factors in the formation of cliffs (vertical scarps up to several metres high).

A recent paper by Julian Orford, which will be published in the proceedings of a conference on shingle beach morphology and ecology, reports on the catastrophic failure and landward migration of an artificially reprofiled gravel beach. Interestingly, the reported retreat rate of the barrier corresponds to 1 m per 1 mm sea level rise. The author did, however, suggest caution in applying this finding elsewhere.

The limited set back of part of the shingle ridge at Salthouse may potentially cause the development of a breach, if the terminus of Blakeney Spit does not migrate landwards. Has the Environment Agency considered this outcome?

S. Hayman, *Paper 7*

The shingle recharge option is one that we exercised a lot of thought over because it could be suggested that in the long term it is more sustainable than the option that we are proposing at the moment, in that eventually, as the shingle ridge retreats it will, in about 100 years' time, overrun the clay bank and then the problem will reoccur. We did not succeed in finding what we considered to be a viable source of shingle. Since the material from the shingle ridge tends to move westwards towards Blakeney Point, there might have been a possibility of recycling some of this material.

However, the whole area is of such importance, environmentally and geomorphologically that we came to the conclusion that the best overall solution was to allow it to revert as far as possible to its natural state and to look towards providing a form of back-up defence. Certainly we were influenced in that decision by the cost implication of the shingle recharge option, which we could safely say would be a factor of 10 greater than the scheme that is being proposed. When we started to look at the proposals, from the point of view of economic viability it was not feasible. During the preparation of the scheme, MAFF decided to take a proactive approach in supporting schemes to protect internationally designated sites, but I think it still behoves us to ensure that the option that we are proposing is the most cost effective, in overall terms, of the ones that are sensibly available to us.

As far as the effect of the management on the shingle ridge is concerned, there is no doubt that latterly, due to the fact that there has been so little available material, we have been steepening the front face beyond that desirable in an attempt to maintain a high crest elevation. This can be counterproductive in accelerating the erosion process and possibly building up a false sense of security in the people who live behind the shingle ridge since they can see the ridge at a certain height but are unaware that it is so vulnerable to attack during storm events. As opposed to gradual flooding resulting from the overtopping of a lower, but more stable ridge, they could be exposed to a sudden breakthrough and inrush of floodwater. We need to be very wary of that situation. There is also a local flood warning system in place as well as the defences which I described.

S. Worrall, *Acting Area Flood Defence Manager, Environment Agency*

The land consolidation schemes used in Denmark are voluntary but backed by the Nature Management Act. How important is legal backing to achieving pilot schemes in the UK and to legitimise transactions of land? For landowners to participate on a large scale, impetus will be needed to start off schemes and to raise the funding to create the land pool.

Dr D.E. Johnson, *Paper 5*
I agree that it is vital to have a proper legislative framework within which voluntary schemes of the sort that I have outlined can work. The joy of the Danish system is that its based on participative management, cooperation and a voluntary ethos within that shell. I think that a pilot scheme could work without legislation but if it were to be taken further then enabling legislation would be needed.

R.G. Purnell, *Paper 23*
The authors of paper 6 suggest a change in the regulatory system; a process that can take many years. I would contend that we all ready have sufficient tools, we are simply not using them effectively. This may be the result of a lack of communication between parties or a lack of understanding of the hopes and aspirations of others.

Delegate
For the past five years, we seem to have been stuck in a bit of loop. We talk, we plan, we develop a shoreline management strategy, and then another, but what actually changes on the ground? It still looks the same to me. I think we face an enormous challenge to get the coast to look as we think it should. We have to undertake something on the scale of what the Victorian engineers did when they first started to engineer the coast, but I cannot see how the current range of legislative discussion is going to deliver the required scale of change. That is why I am raising the question that perhaps there is something legislative standing in the way.

S. John, *Paper 6*
I also think that, before we get to the issue of legislation, I would agree to some extent that there are opportunities within the system but we have to streamline the process. I would refer to the point that we made about a one-stop shop where small developers in particular can call a group of people together, get a decision from them on the type of information to be provided, provide that information and then get a decision on the outcome. The decision might well be negative but the process would have been speeded up, without the necessity of speaking to so many different parties. The potential exists now to do that.

I. Heijne, *WS Atkins Consultants Ltd*
A question for Steve Hayman: could you explain the details of land ownership for the area of retreated land and what would be the implications if it had been farming land?

S. Hayman, *Paper 7*
The marshes were reclaimed in the 17th century for agriculture and most of the land is farmer-owned and used for summer grazing. A large part is owned by Norfolk Wildlife Trust and managed as a bird reserve. We are still in discussion with all of the people with an interest in the land about the details. At the moment it seems quite promising that we will secure landowner agreement, with compensation where appropriate, if we get our planning consent etc. in place. The consultation process started very early in the project and involved everyone with an ownership interest in the marshes. By sitting down and discussing all the different viewpoints we eventually came up with a concensus to which all parties were committed. So from that point of view I would be optimistic that when it comes to agreeing terms we have a scheme that we can proceed with next summer, provided that it satisfies the planning and other requirements.

S. John, *Paper 6*
This is an interesting point and it is an issue that the English Nature and Environment Agency/MAFF project 'Living with the Sea' recognises in that the programme for CHAMPS

and the actual 'Living with the Sea' project have built into them an extensive period for consultation with landowners and monies for acquisition, if that is necessary, all of which is very positive.

Dr J.D. Pos, *Paper 2*
Having being involved with the preparation of the shoreline management plan for the Norfolk coastline, I have a question for Steve Hayman. A partial managed retreat is proposed for the Blakeney frontage. Has the Environment Agency considered the effect of such a partial retreat of this frontage on the frontages to the west of Blakeney Point and whether the policy adopted at Blakeney may lead to a similar policy being adopted for those frontages?

C. Brown, *Seabee Developments*
Steve Hayman makes several comments about the instability of the ridge in the section adjacent to the River Glaven. His overhead slides show a shingle beach overlying a clay bank, with steep slopes at the rear. If the ridge were allowed to roll back naturally, the shingle would, perhaps, block the river, which would cut a new channel to the south. This would allow the more stable cross-section of Blakeney Spit to be formed, which can presumably sustain the phreatic percolation without failure. What would be the objection to making this cut in advance, and establishing the stable profile without the need for an exceptional flood event?

S. Hayman, *Paper 7*
We believe that by allowing the shingle ridge to act in a more natural manner it will roll landwards. For reasons that perhaps are not entirely clear the alignment of the unmanaged ridge to the west has been fairly stable in the past, but that will almost certainly move landwards. The implications as far as the outfall to the Glaven is concerned are being examined. At the moment the Glaven is being squeezed between the shingle ridge and the adjoining area of freshwater marsh, the Blakeney Freshes, so there are major implications for these marshes. These implications are currently being investigated and will be the subject of a follow up environmental statement.

I. Townend, *Paper 12*
My ears pricked up when Professor Hettiarachchi, in this morning's talk about Sri Lanka, said that he concluded an environmental assessment including public consultation in the space of four or five months. In my experience, it can take the same sort of time to get the relevant consenting bodies even to consider the scoping study in this country. So while it may be a matter of contention as to whether we need new legislation or not, it is certainly the case that the present system is creaking at the seams and we need to work together to find a more constructive and streamlined way through the system.

I. Townend, *Paper 12*
Some recent work on volume changes within the Humber estuary identified a similar trend to the mean depths which Professor Reeve showed for the Great Yarmouth banks. However, on closer examination this trend is significantly influenced by the nature of any correction applied to account for the change from lead line to echo survey techniques in the 1930s. Thus, while the techniques described are to be commended as providing an excellent means of exploring the data, it would appear that great care is needed in preparing the data set to be used.
I have a further question for Dr Southgate: when using Monte Carlo simulations it is generally accepted that something in the order of 1000 simulations are needed to obtain

statistically meaningful results. Your examination of wave chronology made use of only 40 runs. How certain are you that you have not sampled a sub-set of the population?

Professor D. Reeve, *Paper 8*

In our study we looked at the sensitivity of the results to the addition of white noise, rather than a step change in measurements, and the results were robust to white noise. However, you are talking about a noise which has, in effect, a distinct spatial/temporal structure and that could have an effect on the average trends that we have found. I think in terms of the overall volume of the sand banks, the figure that we are talking about — half a million — is not a great deal in comparison to something like the 10^9 estimated volume of the sand banks, but perhaps this is an area that warrants further examination.

Dr H.N. Southgate, *Paper 9*

Concerning your question about the number of runs needed to establish the effects of wave chronology, the item on Monte Carlo simulations that I showed referred to an attempt to understand the effects of random errors in the input data. That is a different situation to understanding the effects of different sequences of waves. I fully agree that a very large number of runs are needed in a Monte Carlo simulation to obtain statistically stable results. Some idea of that can be judged from the fact that if you have N variables and P values of each variable then you need P to the power N model runs to completely cover all possible combinations, and you can see that that very rapidly becomes a huge number. For the wave chronology tests, I think that the situation is somewhat different. We have completed some tests with different numbers of re-ordering from very low numbers, 5 or 10, up to over 100 and we found that after about 30 or 40 re-orderings we got statistically stable results. So that is the basis for doing 40 runs.

H. Hall, *Posford Duvivier*

I heartily agree with the speakers regarding the freedom of information and ease of accessibility through web-based systems. What are the 'ethics' for charging for this information?

P. Sayers, *Paper 11*

I think that the key here is that if the information is publicly funded there is a strong argument for that data to be made available freely or just at cost, although obviously there are many commercial organisations which perform value added processing of that data, which is rightly paid for. So I do agree that some information should be freely available, and perhaps we need some kind of national body to put that into practice. However, equally, a lot of information is processed and that will, I think, always fall within the cost domain.

K. Millard, *Papers 10 and 11*

The question of data exchange often arises at conferences and I think it is important to point out that this year the EC launched a Green Paper discussion on the use of publicly funded data, in recognition of the fact that a lot of data collected by public funds is heavily under-utilised. This discussion looked at the various practices in different European countries and found that some countries apply the rule that if information has been gathered through the use of public funds it must be made freely available at cost, as opposed to given away free. However, in the UK there is much more of an onus on organisations to recoup some of their data collection costs and in some cases to offer market prices for the data. The conclusions of the Green Paper were that Europe should move towards a 'norm' of making data available on a marginal cost basis, just covering the cost of providing the data. It was recognised that a certain amount of the funding that public bodies receive does go directly towards delivering certain data at 'no cost' and this should be maintained. Anything beyond that should be examined on a case-by-case basis and a marginal cost process should be adopted.

So although I think it unlikely that we will be able to access all data completely free of charge, there should be more a more consistent approach, the lack of which has posed the major problem in the past. People have been unsure just how much funding should be set aside for data and, from other studies, we have found that in principle people do not object to paying for data, what they do object to is this practice of plucking figures for data costs out of the air. If people know in advance what data is going to cost them, they can set their budgets accordingly.

MANAGEMENT PLANS

Dr D.E. Johnson, *Paper 5*
A question for Martin Mannion: Montrose Bay is important for migrating salmon. Could you comment on the consideration of fisheries' interests within your shoreline management study?

M.B. Mannion, *Paper 13*
We consulted fisheries organisations as we went along. In fact, in the area where the scheme has been carried out there are rights to undertake salmon netting, but these have not been taken up for at least ten years and there is no intention to take them up in the future. So the scheme was not going to disturb netting during construction and, in the long term, the management plan recommendations will not affect the fisheries' interests. The option that the fisheries were most concerned to examine was the proposal to change the practice from dumping material offshore from the port to dumping it in the middle of the bay. However, at the end of the day, they were quite happy to accept that change of practice.

T. Collins, English Nature
Relocation of parts of the golf course at Montrose rather than its protection via coastal defence measures is an admirable solution. Does the speaker wish to comment on the application of this approach elsewhere in the UK where, presumably, the funds available for coastal defence measures are not available to fund the cheaper, more sustainable and more environmentally acceptable alternative option.

M.B. Mannion, *Paper 13*
It is certainly a dilemma. The circumstances just happened to be right in Montrose. Relocating the golf course does not, of course, attract coast protection funding but it was an option that the council were willing to accept. The course is leased out to a club and they also agreed to the proposal, realising that they could not stand up against nature. But while the situation was right in Montrose to make that decision, it would not always be appropriate to adopt this course of action elsewhere and there is no fixed mechanism.

Delegate
I have to make a point for MAFF here. The question really depends on who is going to pay for the solution. There is always this difficulty of whether the taxpayer should step in and bail the landowner out of a difficult situation. Each situation must be looked at from a national point of view. I think the situation was easier in Montrose Bay because the land was owned by the council.

M.B Mannion, *Paper 13*
That did make matters easier. Early and ongoing consultation, involvement of key stakeholders and willingness to examine wider considerations also helped. We need to be open to looking beyond 'do nothing' to the coastline in coastal defence terms.

J. Brooke, *Paper 17*
In coastal defence terms, does the duty of care apply differently in those situations where landfill or similar materials would be exposed by erosion when compared to situations where soils, clays etc. would be exposed?

D.J. McGlashan, *Paper 14*
We have not actually seen any cases in court, but my guess would be that extra problems would arise in the case of a cliff that is contaminated with some special waste, perhaps even issues relating to who dumped the waste in the first place and possibly the Pollution Control Act would be involved. I am not entirely sure. This is a question which requires further investigation.

C. Brown, *Seabee Developments*
Has Derek McGlashan looked for comparisons with law in other English speaking countries, e.g. Australia, USA. I had experience where B sued planners for consenting to a structure. After seven years this case reached the Supreme Court, which found for the planner, but with the proviso that the finding was due to the lack of knowledge of the original planning consent at the time. Any future case would be heard on the basis that the knowledge exists. Secondly, if B is followed by C, D etc. and the planning authority agrees that the scheme is sound and necessary, should B be able to prevent it? A comparison may be made with the situation for Flood Prevention Associations, whereby membership is compulsory if 90% of the people who are subject to the flooding are in favour of a scheme.

D.J. McGlashan, *Paper 14*
It is difficult to access the legal system in other continents so, to date, it has been a matter of focusing on the differences between Scottish and English law and the applicability to issues in the UK. However, over the next three years we hope to examine international examples.

Professor C. Fleming, *Halcrow Group Limited*
Shoreline Management Plans attempt to predict where the shoreline will be over a seventy-five year horizon. These estimates also consider the impact of sea level rise. What do you do within an estuarine environment with respect to predicting or estimating similar long-term morphological development?

Dr N.I. Pontee, *Paper 12*
We had in mind an estuary-wide approach, and I also briefly mentioned the roll-over model for estuary evolution, which is similar to the evolution of a coast in that the beach profile moves landwards and upwards with sea level rise. Our approach was to extend that model to an estuary. John Pethick was the proponent of the model which basically involves the whole estuary system moving upwards and landwards with sea level rise. You can see this process in the Severn where the seaward extent of the estuary widens and erodes and then you make accretion for the landwards. That was the sort of model that we had in mind for the evolution of the estuary system.

I. Townend, *Paper 12*
May I make a further comment on Dr Pontee's answer to Chris Fleming. I think we have to accept that within an estuary situation we are at a fairly primitive stage in terms of predicting long-term change. I certainly would not wish to have to predict what an estuary will look like in 75 years' time. However, a fairly major research initiative has been started, funded by the Environment Agency, MAFF and English Nature, and similar initiatives are being started — NERC and EPSRC. So work is beginning on trying to determine how an estuary will evolve in response to such matters as sea level rise, which is a very difficult question because it is a matter of disentangling a closed loop. The shape of the estuary is dependent on the hydraulics which are in turn dependent on the shape and actually finding a way to break into that loop poses quite a difficult problem which will be the focus of this research programme. So while we do have hypotheses and concepts that allow us to make estimates which provide a starting point for current estuary shoreline management plans, we must accept that there is a long way to go on this issue.

Dr D. Brook, *HOB Planning and the Physical Environment, DETR*
Just a brief comment on the liabilities of planning authorities in granting planning permission for coastal defence works. English case law determined that local planning authorities do not have a duty of care to individuals in granting permission. The duty is to the public interest and it is accepted that a permission may be to the benefit of the public as a whole while adversely affecting the individual.

My other point was a quick question to Martin Mannion in relation to the Glaxo Welcome plant. Who actually funded the works?

M.B. Mannion, *Paper 13*
Glaxo Welcome funded the scheme in front of their site. They are the principal landowner and they were prepared to provide the funding. The Angus Council will undertake the other intervention options within the bay.

J. Riby, *Paper 16*
May I just make two points about the Holbeck case without compromising the Council's position in the appeal. Judge Hicks actually found against us [Scarborough Borough Council], not as a coastal authority or a council, but as a landowner. As the matter stands at the moment, however, we are hopeful. The other point I would like to make is that when people are carrying out work and have information available they should consider passing that information on to third party landowners, etc. As a public body we conduct studies and have surveys carried out, and sometimes that is where the matter stops; it is purely a matter of providing useful background information. However, if you have information that may have an impact on a third party you should really consider whether to pass on that information.

R.G. Purnell, *Paper 23*
An interesting point worthy of further consideration. The production of Shoreline Management Plans has provided extended information on flood risk and erosion to which some authorities want to restrict access. Fear of causing planning blight is the usual reason.

W. Brook-Hart, *Gifford & Partners*
A question for Jan Brooke: given that different interest groups tend to have different amounts of time available to attend seminars and consultation forums, do you need to take any special measures to ensure that a reasonable balance of interests are represented at your seminars?

J. Brooke, *Paper 17*

One particular step that we took was the use of newsletters to give people plenty of warning of the consultation seminars, wherever possible two or three months in advance. This is not always easy within a tight time-scale dictated by a defence project. Another option which we proposed but did not actually follow through was the opportunity to attend an evening meeting. We felt that some of the NGO representatives with full-time day jobs might want to participate in the evenings, but in each case we had so few responses to the evening meeting option and those that did wish to attend were, in fact, also able to attend afternoon sessions, that the idea was effectively redundant. However, I think the important issue is to offer people the opportunity. It is better to anticipate a need that may turn out not to be necessary than to risk the criticism that you are excluding people. So many of the steps were preemptive and that, coupled with the use of newsletters (rather than letters which require a response) simply appraising people of the situation but not asking them to comment or actually to do anything, I believe has also helped us to reach a wider audience.

J. Pos, *Paper 2*

Regarding consultation, it has been our experience with shoreline management plans, and I am sure that of every consulting engineer in this room, that invariably the client underestimates the amount of time that is required for consultation. In a shoreline management plan which we had prepared, the hours allocated to consultation in the brief were far exceeded because, inevitably, a cascade situation develops. People that you target specifically as having an interest in a particular issue then tell you of other people who have an interest and so on. So I would say, in preparing briefs, be generous with the time that you allow for consultation.

D. Swan, *Dean and Dyball Ltd*

A question for John Riby: having finalised your plan you mentioned looking at different procurement routes. Could you expand on that?

J. Riby, *Paper 16*

We have already made a joint bid with MAFF to the Treasury under the modernisation capital fund. Unfortunately, the first round of results did not favour our bid. We are now discussing with MAFF other means of procuring funding for, at the very least, an investigation into how we might deliver coast protection. There has been some salutary experience with people spending an awful lot of money putting forward cases for PFI type schemes only to find that the schemes did not find favour; a risky business when you are working with limited capital resources. However, we believe there may be a way of resolving this issue jointly, and it will be to the benefit of the industry in general if Scarborough is able to work out something with MAFF, then we can share that knowledge with others because I am convinced that this is the way that things will go in the future.

Dr D. Brook, *HOB Planning and the Physical Environment, DETR*

Looking beyond Scarborough Town, where the SMP policy of holding the line makes life a lot easier, have there been any difficulties in integrating the outcome of the SMP with the strategy planning system, particularly in areas where the policy might be retreat or do nothing?

J. Riby, *Paper 16*

I am very fortunate to be involved in a group where the planning authorities are definitely onside and they are actually endorsing a lot of what we are trying to achieve in the shoreline management plan. So the succinct answer is that we do not appear to have a problem. The

National Park is very supportive and a lot of our frontage within the National Park does, of course, fall within either the do nothing or managed retreat approaches so we are not actually advocating human intervention. But, interestingly, in the shoreline management plan we have asked parties such as the National Park to do some research themselves and they are actually bound into the agreement, so it is not just operating authorities that are carrying out these studies, other parties are expected to do them as well.

S. John, *Paper 6*
The integral approach taken in the Suffolk Estuaries Strategy to allow systems to function effectively is very positive. However, could Nick Pettitt please expand on the conservation bodies' view on relocating the SPA within the Blyth Estuary to create a system that functions more effectively; and will the Habitats Directive allow such flexibility to create a better system?

N. Pettitt, *Paper 15*
On the Blyth Estuary, the Tinkers Marsh SPA is one of our biggest concerns. In an ideal situation we would want to hold the line there to continue to protect the marsh but we are trying to look into the future, to assess what flood defence attitudes might be in say 20 or 30 years' time: then people might very well say 'why on earth are you defending that freshwater marsh below sea level?' We have also had several in-depth discussions with RSPB and English Nature and the Suffolk Wildlife Trust, mainly on the subject of how and to what degree to implement the European Habitats Directive. I think there is a general recognition between the bodies which look after the habitats around the Suffolk and Blyth Estuaries that something needs to be done, and the most practical solution would be to relocate Tinkers Marsh. Much more work has to be done to determine exactly how to implement the Habitats Directive. I do not think that Tinkers Marsh is a unique case, this will apply to many estuaries and areas of coastline. So this is not a problem specific to the Suffolk Estuaries but rather that the interpretation of the Habitats Directive in general has to be decided and agreed upon.

I. Townend, *Paper 12*
Given the large scale changes and removal of defences being proposed along some reaches of the Blyth, coupled with continued constraint of others, what geomorphological response do you anticipate?

N. Pettitt, *Paper 15*
I would not say that we are proposing large scale removal of defences. The discussion following the first session about predicting the geomorphology and the evolution of estuaries in the future seemed to result in the opinion that it was very difficult to do. There are extensive flood defences throughout all three estuaries, but on the Blyth in particular, which do tend to override the natural evolution of the estuary. So as well as considering the natural evolution we also have to take a very detailed look at how each length of defence is currently performing and what it is capable of doing. The present defences are the key factor in predicting what is going to happen next, outweighing the influence of geomorphology.

D. Hattersley, *Charles Brand Limited*
A question for John Riby. Regarding a comment that John made about the Local Government Association having a special group working on coastal strategy. Could he expand on that and say whether this is an extension of the Shoreline Management Plans or a completely different tack, perhaps based on not wanting to have a 'Holbeck Hall' on their doorstep.

J. Riby, *Paper 16*

The Local Government Association (LGA) Special Interest Group on Coastal Issues is an elected member led group looking at all the different aspects of the coast. The LGA is very concerned that the coastal margin is not getting its fair share of financial commitment from Central Government. For example, many British seaside resorts have an artificial 'veneer' that people see when they visit throughout the summer. What they do not see during the winter is the deprivation, unemployment and things like the conversion of hotels/guest houses into flats and multiple occupancy accommodation, etc. The LGA Special Interest Group on Coastal Issues was set up with its primary concern being to the coastal zone in all its facets.

Dr S.S.L. Hettiarachchi, *Paper 1*

We have heard a lot about public consultation and I would like to share a view of public consultation from the developing world. In most projects in Sri Lanka, as well as in other parts of the developing world, public consultation is a very difficult issue because you are dealing with such a varied cross-section of society. The leadership provided by pressure groups, NGOs and other interested parties can produce a major force within the public, although the people themselves may not actually be very well informed on the particular subject. This can cause major problems.

J. Brown, *Severn Estuary Strategy*

Jan Brooke and her colleagues are to be congratulated on their efforts to ensure transparency, dissemination of information and wide consultation on the Bristol Channel project while work has been underway. I wished to know if there are any mechanisms in place to aid the availability of the extensive data gathered during the life of the project, in order to make best use of it in the future, with the proviso that some of the information is of commercial value and was submitted in confidence. The Severn Estuary Strategy (SES) has informed many of our partners about the study. Data collected for the Bristol Channel project would be of great value to the SES and many of our partner organisations.

J. Brooke, *Paper 17*

This question also relates to some of the papers that were given yesterday on the availability of data. In a previous discussion, the same question was raised about whether the data that had been collected on a GIS would be available, and the response from the National Assembly for Wales was that they were aware that groups and individuals had contributed data in good faith, that they were interested in having access to the database, and that they (NAW) would try to ensure that it was possible. Clearly this raises the issue of ensuring that the available data is up to date, so the National Assembly for Wales are facing a challenge, having been provided with a system and a method, on how best to keep it running and relevant. It was encouraging to see a recognition of the need to maintain the goodwill that has been generated by those people who have offered data, and to make that data available to interested parties through some means in the future.

RISK

D. Ayers, *Ministry of Agriculture, Fisheries and Food, Taunton Region*
I would like to congratulate Dr West and his team and all those others involved in the development and construction of the Minehead project. It has been an exceptional example of teamwork and partnering by consent, and the result has been praise from locals and tourists alike. It is a real success on virtually all fronts. But before anyone leaves the conference with the impression that deferred payment is the answer to funding problems I would like to add a cautionary word. Firstly, the financing approach was not approved or agreed by MAFF, as stated in the paper. Deferred payments are not acceptable and are seen as a means of circumventing the controls of government accounting procedures. For this reason Treasury agreement was not forthcoming retrospectively and, moreover, it was stated that agreement would not have been given in advance. The possibility of qualification of accounts and financial penalties is another risk that needs to be addressed, particularly for novel procurement methods.

Dr H. Southgate, *Paper 9*
A question for Jim Hall: when engineering consultants buy a mathematical model, they like to have the assurance that the model has been validated. Validation is difficult to define. It means different things in the software industry, academia and engineering design. How do you relate your concept of model dependability to model validation?

J.W. Hall, *Paper 19*
Validation is indeed a thorny issue, but considered in its most general sense it is to do with collecting evidence which supports one's belief in a model, so in all of our model studies we should be looking at ways of structuring that evidence logically to support the validity of our models. I believe that the reason why establishing validity is problematic is because sometimes people connect it with the idea of the model being in some sense 'true'. Obviously, all of our models are merely abstractions of reality, so we should be collecting evidence which describes what that abstraction is and then trying to present that evidence in a logical way.

K. Millard, *Papers 10 and 11*
Jim Hall introduced the concept of model 'dependability' in risk evaluation. How is the concept of 'dependability' related to more established concepts such as 'accuracy' ?

J.W. Hall, *Paper 19*
I think accuracy can be connected with the ideas of truth and precision, whereas if one is considering dependability one is approaching the concept from the opposite direction, working backwards from a premise and asking 'does this model provide me with the information I need in order to make my decision?' If the model does provide a useful basis for the decision making then I would consider it to be dependable. 'Accuracy' perhaps considers the model in a rather more constrained context, without looking at the purpose for which it is being used.

D. Shercliff, *Geofabrics Ltd*
It is very encouraging to hear that coastal defence projects are being designed using risk based analysis and that each component part of the construction is given a risk value. Dr West mentioned that inclusion of rock work in phase 1 would reduce the risk of non-performance (i.e. overtopping) and I am interested to know if the risk assessment extended to comparison of revetment systems and the inclusion or omission of a geotextile in the

construction. The Rock element of a revetment system is an expensive item which needs to be cost-justified as well as design-justified. However, the geotextile is a relatively low cost component but its performance is critical. Proper specification, procurement, quality assurance and installation of the geotextile all have a bearing on the ultimate function of the embankment. If a risk value were placed on the geotextile, an engineer faced with a lesser alternative would be able to assess the consequences more easily and ensure 'best value' for the client.

Extensive testing of various flexible revetments has been carried out on Dutch coastlines (reported in ICE conference proceedings). Comparisons of revetments and inclusion or omission of various geotextiles have been made, producing the statistical evidence required for placing a risk value on the component parts of the revetment. The experience in one location was 'no geotextile — no embankment' following a storm event shortly after construction.

What risk value did you put on the use or omission of a geotextile on the Minehead sea defence scheme?

Dr M.S. West, *Paper 20*
The paper that we presented focused mainly on the Phase 2 works, the beach nourishment, and it would be fair to say that most of the risk assessment and the management of risk and uncertainty in the paper has been concerned with Phase 2. Regarding Phase 1, it must be remembered that at Minehead we have a sea wall — the wave return wall — at the rear of the defence, running along the whole length. It changes in height and cross-section as it moves from the more exposed to the less exposed regions, but there is always a backstop — the concrete wave return wall behind a concrete step revetment along Warren Road. So the short answer is 'no', the use of geotextiles did not enter the risk assessment.

T. Collins, *English Nature*
Regarding the development of the shingle recycling at Dungeness: could Dr Maddrell tell us something about the management of risk in future? Presumably the site will need protection for hundreds of years, the coastline west of the site will retreat, become swash aligned and longshore drift from this area across the power station frontage will decrease, with inevitable implications for the costs and sustainability of the recycling regime. What are the current plans to manage this site in the long term?

Dr R.J. Maddrell, *Paper 18*
These stations are not designed to be productive even another 50 years, although the Magnox, which should have been decommissioned about 5 years ago, still keeps going. Clearly the amounts of material being moved are going to increase in the future. That cannot be changed, but in the long term, perhaps when sites are decommissioned, some erosion and managed retreat may have to be allowed up to a particular point. It seems sensible to make use of what you have. But in the end, in say 100 years' time, you may actually have to go and remove whatever of the Stations you have left behind, the concrete of the core and any contaminated matter, and take it away.

J. Rawson, *Environment Agency*
With respect to the technical success of any engineering project, the associated risks may be reduced if we ensure that as many lessons as possible are learned from previous projects. Indeed, many speakers have referred to the importance of monitoring the performance of schemes, partly for this reason. An example where this approach has been taken is in the Wash realignment mentioned by Dr Maddrell, where lessons from Tollesbury and Orplands have been fed into the scheme design.

In its guidance, MAFF supports the evaluation of the success of a scheme in achieving its design aims. However, the majority of Post-Project Evaluations (PPEs) completed to date concentrate solely on adherence to budget and time-frame. Do the speakers have any suggestions as to how geomorphological and other technical considerations could best be included in PPEs? Could a standard approach be recommended? I would also welcome comments from MAFF regarding funding for such evaluations, possibly as part of overall scheme costs, and any news on the forthcoming PAGN on this topic?

Dr R.J. Maddrell, *Paper 18*
I think that an understanding of the recent geological history, i.e. the geomorphology is absolutely critical and I believe that such research should be funded. Unless you have an understanding of the past, you cannot apply even a short-term policy let alone a long-term policy of what to do with a power station once it has been decommissioned.

R.G. Purnell, *Paper 23*
It is not true to say that post-project evaluations concentrated on the financial element. There were two types of post-project evaluation: one looked at the financial side and how it was constructed, and the other how the scheme performed. However, although we invested a lot of money, we did not learn a lot. Our aim now is to produce new project appraisal guidance on post-project evaluation but, to be honest, we are not entirely sure how to maximise the benefit of PPE. As far as funding is concerned, obviously no public authority would start a project without first considering post-project evaluation, whether funding was coming from MAFF or not.

P. Sayers, *Paper 11*
We have been in discussion regarding the policy aspects of nuclear power at Dungeness, and their decommissioning plans are largely what they call a long-term safe store, which involves removing as much material as possible and then, 127 years after cessation of power generation, infilling with concrete. So the site is there for a very long time and will always have to be managed.

Jim Hall raised an interesting point regarding acceptable risk levels being applied to coastal engineering, drawing comparisons with, say, the dam industry which has a particular standard that they have to achieve, a probability of failure of 10-6 or so, but that is related to loss of life and the safety issue. In coastal defence appraisals where we are considering safety, are we considering loss of life?

J.W. Hall, *Paper 19*
Flood defence objectives are quite clear and deal with reducing risk to people, property and the natural environment. So we have a range of different objectives and on the whole we tend to cluster them all into one economic measure. One message to come out of the Bye Report was the vital importance of the health impacts, the loss of life impacts and also the social impacts of flooding, so clearly these are the priorities and it seems as though some of those aspects may not be fully reflected in our economic appraisals. So in those cases where there is potential for devastating damage, we should be trying to collect evidence about acceptability from as many sources as possible. The economic appraisal will be one of the key sources of evidence, but because the impacts of these devastating floods are so uncertain, I think it is reasonable to look elsewhere for evidence to help us in our decision making process.

J. Cocker, *Teignbridge District Council*

Coastal defence is integrated with land use and planning development, together with economic and leisure strategies. I believe that local authorities balance those interests rather well at the moment. So why does Dr Maddrell think that a national flood and coastal defence authority is required?

Dr R.J. Maddrell, *Paper 18*

Basically, I believe that the whole process might be more efficient under the auspices of a single authority. I am not saying that the current system is inefficient (and the local authorities make a very useful contribution), but it may be more efficient to have just one body in charge. It may also be that having one agency, it might be able to bypass the political wrangling, avoiding situations such as in the aftermath of the 1953 flood where the political requirement was not simply to hold the line, but to reclaim land if at all possible. We are now having to live with and pay for the consequences of this earlier policy.

INTO THE NEXT MILLENNIUM

J. Brown, *Severn Estuary Strategy*

Heidi Roberts stated in her presentation (Paper 21) that 'the SMP process was invaluable in bringing together planners, engineers, ecologists and archeologists ...'. I am not convinced that this is always the case, so this is a plea to those of you who are engineers in local authorities, to ensure that your planning colleagues are well informed about the SMP, and to ensure that you communicate adequately within your organisations.

The value of estuary and coastal projects is generally accepted in the UK, but an accepted methodology to evaluate estuary/coastal projects is needed, particularly for the evaluation of implementation, to prove the value of projects.

Political will at a national and regional government level is required to provide better support for holistic management of all our coasts and estuaries in the long term. There is currently a vacuum at national level. Coastal practitioners, particularly those involved in voluntary initiatives, have been making this same complaint for the last few years. I have heard this issue raised again and again at conferences and workshops, yet it seems to no avail. I fear this is an issue low on the political agenda. If guidance does not come from Europe in the immediate future, then support for sustainable projects at a national level is a more immediate issue. Coastal and estuary practitioners are aware of the need for integrated management of the coast, some have knowledge or experience of integrated coastal zone management (ICZM) in other countries around the world. So should coastal and estuary practitioners in the UK formulate a proposal for holistic and integrated management of the whole UK coastline and all estuaries, to submit to the UK government?

H. Roberts, *Paper 21*

In answer to your comment on communication between departments within organisations, I agree totally. One of the main themes that we examined within the demonstration programme is information dissemination, and we found this important issue within that broader issue. People tend to focus on information exchange between organisations but it is just as important, or even more important, within the departments of the organisation itself. That is a problem that still needs to be addressed within each organisation.

The establishment of an accepted methodology to demonstrate the benefits of implementation is needed. It is very difficult to actually assess how well we are doing, and people continually ask about the benefits of estuary management and the benefits of actually implementing these various policies. It is an area which requires much more research in the future, and hopefully European funding will eventually be available for that.

With regard to whether we should take the bottom up approach to coastal zone management, I think there should be a combination of both. We can and should do all we can from the bottom up, and I think that lobbying the DETR and the European Union as often as possible is helpful, but unless we actually get some support from central government then we will continually be explaining to people what we are doing, and justifying the benefits. So without government support it is going to be difficult to prove the legitimacy of coastal zone management.

R.G. Purnell, *Paper 23*

Whenever I hear calls for CZM legislation or further advice I tend to wonder whether it is simply an excuse for lethargy as the procedures requested are already in place. We already have a structure planning process. If I were a planner putting together a structure plan I would have to take into account everything that would have an impact on the area. I would want to incorporate transport policies, coastal policies, etc. How could I produce a structure plan without examining all those aspects. So why do we need Europe to tell us how to produce integrated coastal zone management when we already have the procedures in place?

Dr D. Brook, *HOB Planning and the Physical Environment, DETR*

I do hear the calls for something to be done about the policy vacuum and the lack guidance from central government, but I disagree. I think the guidance is there. The message that is coming through is that you are not actually asking for guidance but for money, and there is a difference between the two. However, if you really do need guidance then please tell us what form of guidance you need. Get the LGA together to produce their coastal strategy, submit it to central government and explain exactly what is needed with regard to the coastline. The Resorts Association have done just that. They produced a small document on British Seaside Resorts, an excellent publication which identifies the problems and defines a methodology for moving forward. That methodology has been accepted by CMS and included in the CMS Tourism Strategy. If you work from the bottom up level and present your strategies to central government, central government will include them.

H. Roberts, *Paper 21*

There is indeed guidance available but it is mainly focused on shoreline management and planning policy on the coast and we are hoping for guidance on the wider aspects, to incorporate all the different sectors of the community. Having undertaken many different initiatives and testing various methodologies we would appreciate some feedback from government to let us know if we are heading the right way and if anything ought to be done differently.

Dr J.D. Pos, *Paper 2*

Should we not explore better ways of informing the statutory structure and local plans developed by local authorities to move towards ICZM. From my experience on the Essex SMP, extensive consultation with the local planning authorities helped to define sustainable coastal zone policies both for the SMP and for future revisions of the statutory plans.

H. Roberts, *Paper 21*

The shoreline management plans are constantly evolving and they are now starting to take wider aspects on board, so things are moving in the right direction. We are examining the broader implications of shoreline management and, hopefully, in the next round of structure plans etc., coupled with the local agenda 21 plans, we will start to look holistically at every plan that we produce.

J. Brooke, *Paper17*
A couple of important points have been raised at this afternoon's session: the importance of the role of planners in the SMP process, in the planning system for ICZM, in tourism initiatives, and we have just heard a plea to involve planners in coastal management generally. Perhaps there is a challenge here for the Maritime Board for the next conference. The other point concerns target setting, and I was just wondering how Reg Purnell equates this 'leap forward' with two steps back.

R.G. Purnell, *Paper 23*
The targets will hopefully be published within the next few weeks, but it must be an on-going process. With the introduction of OPMs and targets we are continually trying to improve the decision making process and build on what we have done before. Naturally, not all of our past attempts have been successful but that should not stop us trying.

Captain A. Wilks, *Scottish Coastal and Forth Estuary Forum*
As MAFF is the principal funder of coastal protection but ICZM comes under the DETR, would there be a benefit to the coast if each could be centralised in one government department?

R.G. Purnell, *Paper 23*
The Government has already answered that question in its response to the Agriculture Select Committee Report. It said that responsibilities will stay where they are. In dealing with government issues, this difficult decision of deciding where boundaries will be drawn always arises. If you extend the argument you could put the case that parts of the DETR should be within the Treasury or Home Office, and then real complications would arise. It is up to us to make the current situation work. The Prime Minister recently made an interesting comment on the concept of joined up government. That is where our responsibility lies; we have to liaise with each other, and sometimes we even agree!